Silk Road of Flow Technologies

Photo 1.1 Nilometer at Aswan of Nile River

Photo 1.10 Hummer Noria

Photo 1.5 Drainage outlet of Cloaca Maxima, recently

Photo 2.11 Vertical windmill in Iran

Photo 2.13 Classic windmill in Zaanse Schans near Amsterdam

Fig. 3.9 Many Archimedean spiral pumps used in "Sado Kinzan Kanahori no Zu" in Edo era (Cabinet Library Collection)

Photo 3.16 Japan first locomotive From Shimbashi to Yokohama

Fig. 5.59 Appearance of female driver (early Showa period)

Photo 6.1 MacArthur arrived at Atsugi Naval Airfield on the Bataan. He is the second from the right

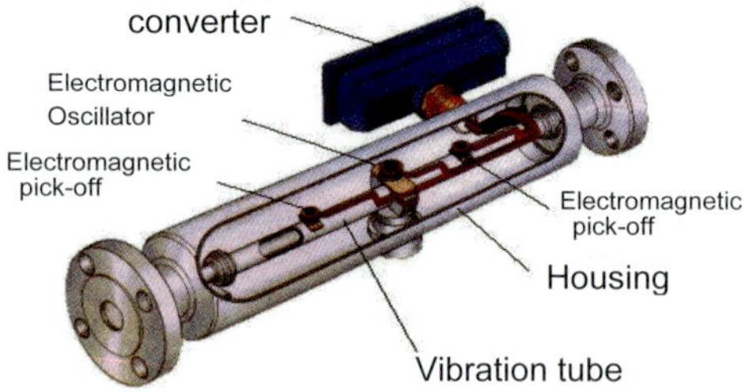

Fig. 6.38 One -straight tube type Coriolis flowmeter of Oval

Photo 6.27 Japanese weighing oil flow standard

Actual Flow Facility for High Reynolds Number in Japan.

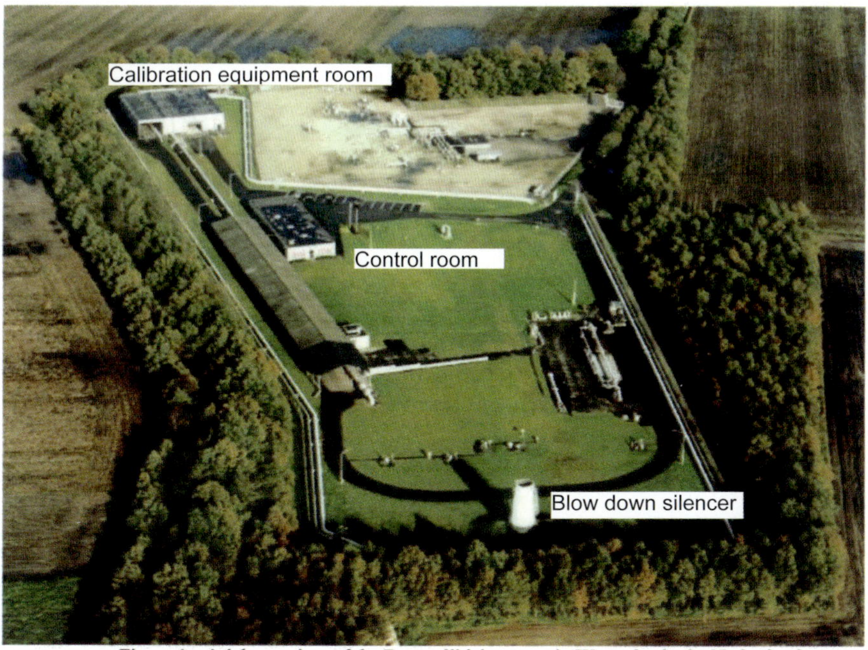

Photo 6.29 Bernoulli laboratory in Westerbork

Silk Road of Flow Technologies

The Japanese Edition Published in 2006 by JAPAN INDUSTRIAL PUBLISHING CO., LTD.

ISBN: 978-1-64273-322-8

Written by Yutaka Ogawa
Projct Editor: Yumi Itabashi, Hachidori Inc.
Editor: Andrew Kirkham

Printed in Japan
1 2 3 4 5 6 7 8 9 10

One Peace Books
43-32 22nd Street STE 204 Long Island City New York 11101
www.onepeacebooks.com

Photo 6.27 Japanese weighing oil flow standard

Actual Flow Facility for High Reynolds Number in Japan.

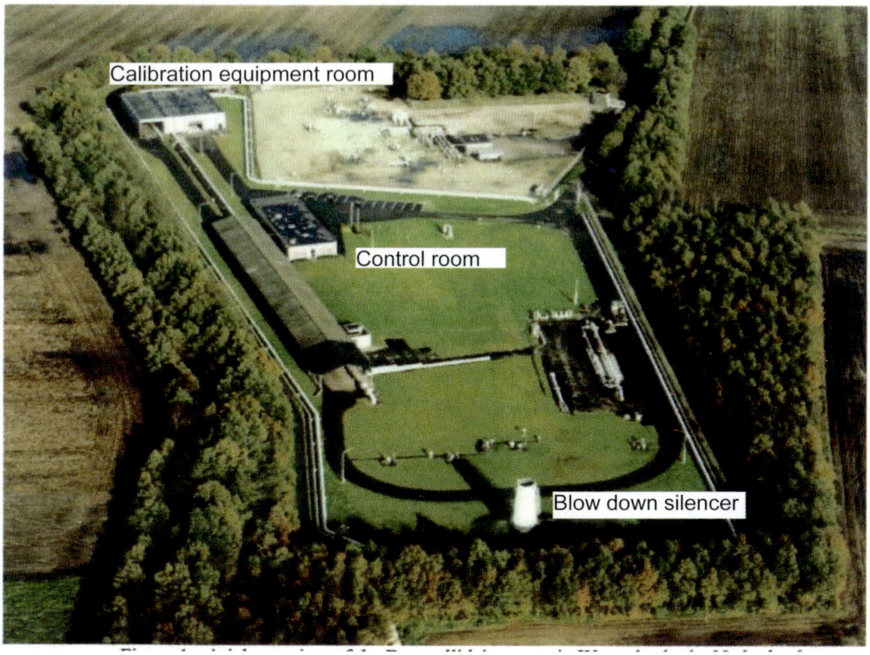

Photo 6.29 Bernoulli laboratory in Westerbork

Silk Road of Flow Technologies

The Japanese Edition Published in 2006 by JAPAN INDUSTRIAL PUBLISHING CO., LTD.

ISBN: 978-1-64273-322-8

Written by Yutaka Ogawa
Projct Editor: Yumi Itabashi, Hachidori Inc.
Editor: Andrew Kirkham

Printed in Japan
1 2 3 4 5 6 7 8 9 10

One Peace Books
43-32 22nd Street STE 204 Long Island City New York 11101
www.onepeacebooks.com

CONTENTS

Silk Road
of
Flow Technologies

Yutaka Ogawa

1. Relationship Between Water and Humankind

1. 1 Ancient Orient

1.1.1 Origin of the Water Level Gauge [1]

Tracing histories of humankind, in the primitive era, it was necessary to secure food by hunting and to secure drinking a water from rivers and lakes for life. Irrigation would become important when plant foods were cultivated. In a water-rich country like Japan, it would not have been so difficult to secure water, but on the African continent where the rainy season and dry season alternated, there would be the progress in the development of a water storage technology like irrigation along with the fight against water. It is not hard to imagine that technologies for transporting drinking water and irrigation water from distances would have developed in desert and dry areas.

The Nile basin in Egypt has been a unique region as one of the very fertile lands in spite of no rainfall. As Herodotus said, it is truly a "gift of the Nile."

King Akhenaten praised the Sun God, "You have created the Nile in lower world, and by your will, you have brought the Nile to give vigor to the Egyptian people."

Successive pharaohs were guardians and promoters of irrigation and farming. Amenemhat I of the 12th Dynasty (c. 2000 BC) said, "I have grown grain and loved the grain-god Neper. The Nile has welcomed me in every basin and no one has been crying and suffering from thirst." However, the Nile sometimes caused enormous damage to human lives, crops and livestock, as seen in the great flood of the 1874 BC.

For this reason, at least 20 liquid level gauges for observing and recording the water level were attached to the quay of the Nile and the outer wall of the temple standing at intervals along the river. It is the etymology of Nilometer (Nile → Nilo) that is still used in harbors. It is said that the

Photo 1.1 Nilometer at Aswan of Nile River (10)

Nilometer which must have existed as long as mankind had the ability to invent letters and to keep records. Photo 1.1 is a Nilometer that still exists on the Nile. Water level was measured by horizontal lines as seen in the photo. 1.1. The highest flood levels were recorded annually in the documents of palaces and temples and the official chronicles of the Egyptian kings contain numerous records of the Nile water level.

Fig. 1.1 shows changes in the water level of the Nile over 4,000 years, recorded by a Nilometer from the Qarun Lake. Nile River is still equipped with Nilometers, and by comparisons with these records it is shown today that the Nile has risen 6 meters above the sea level according to water-carried sedimentation during the last 5,000 years. Thus, from 3,000 BC, various technologies have emerged in Mesopotamia and Egypt as the result of agricultural

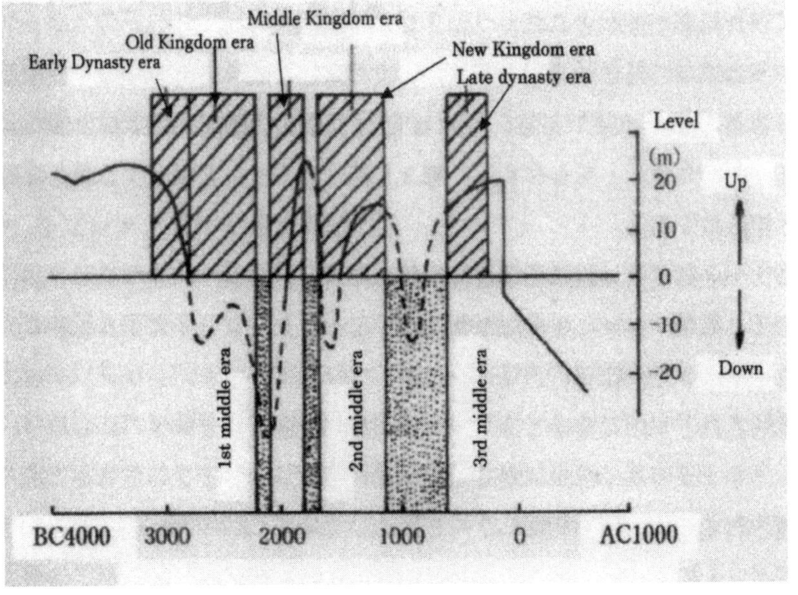

Fig1.1 Change in water level of of the Lake Qarun by Nilometer [11]

revolutions. First of all, canals and artificial irrigations are listed, next, water wheels and plows that were invented in order to use water and animals as power.

The copper water pipe totaling 400 meters length running to a temple from around 2,500 BC was excavated from a Memphis ruin. This was forged copper pipes measuring 4 cm in diameter and 1mm thickness. Also, a culvert of clay pipes from around 1,900 BC can be seen.

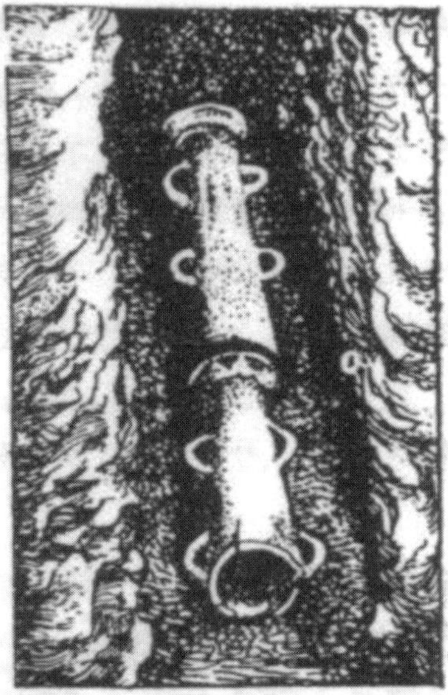

Photo.1.2 Earthenware Water Pipes I Knosos Place

Photo 1.2 shows a contemporary water pipe made from earthenware, which was excavated at the Knossos Palace in Crete. The pipe gradually narrowed and the ring pipe was hardened with binder to create a waterproof joint. It was also designed so that ropes could be passed through the holes in handles on the outside of the pipes in order for each to connect tightly using tension.

1.1.2 Origin of Water-Supply [(1)(4)]

Fig.1.2 illustrates a modern Qanat of Iran. A qanat is a long, sloping tunnel that sends cool, salt-free water from a source of water to arid areas. As shown in the figure, there were vertical shafts about every 45m to send air to the people digging the tunnel and for hoisting out debris.

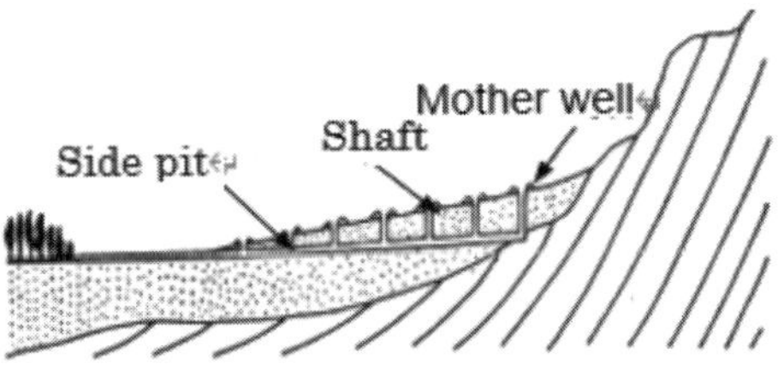

Fig.1.2 Modern Qanat at Iran district (1)

The qanats excavated on the Iranian Plateau have a horizontal hole length of 5 to 10 km and a vertical hole depth of several tens of meters, and there are culverts that allow people to bend down and walk. There is also a large-scale qanat with a total length of 70 km and a depth of 300 m at the deepest part of the hole. Fig.1.3 shows the layout of the craftsmen who excavated the qanat. After the rapid spread of qanats on the Iranian Plateau, the technology of drilling waterways was introduced to West Asia, the Arabian Peninsula, and Greece.

A: Dig the soil with a pickax. It is Determines the excavation direction and the slop of channel.
B: Collect earth and sand, and pack it in a leather bag.
C: Carry a leather bag under a shaft and tie to a hoist,
D: Operate the hoist and pull up the leather bag.
E: Discard the earth and the sand in the leather bag and tie an empty bag to the hoist

Fig.1.3 Layout of craftsmen digging a qanat

With the expansion of the Islamic world, qanats spread to

North Africa and were excavated in Libya, Algeria and Tunisia, and eventually to Europe via Morocco and the Canary Islands, and Bohemia in Spain, Germany's Bavarian and Weimar regions, and Czechoslovakia. In addition to being established in rural areas, there is a theory that the underground waterways in Mexico, Peru, and Chile in Central and South America were brought by the Spanish.

Meanwhile, they also exist in Afghanistan, Pakistan, the North Indian region, and republics near Central Asia and Russia. In China, it is stated in the historical "River & Canal Book" that a qanat was dug in Chen prefecture, Shaanxi province in the Han dynasty.

It is thought that they were also conveyed to the east via the Silk Road. In Japan, there is an underground irrigation route called "Mambo" that straddles Mie and Gifu prefectures. Since this pronunciation is very similar to the existing "Mansunbo" in South Korea, it can be imagined that the qanat technology was been passed down to Japan through the Korean Peninsula from China.

I went to see Mambo in Japan in 2008, so would like to describe it. (See Fig.1.4)

In Japan, there are "Manbos" at Katahi aria Daian-cho,

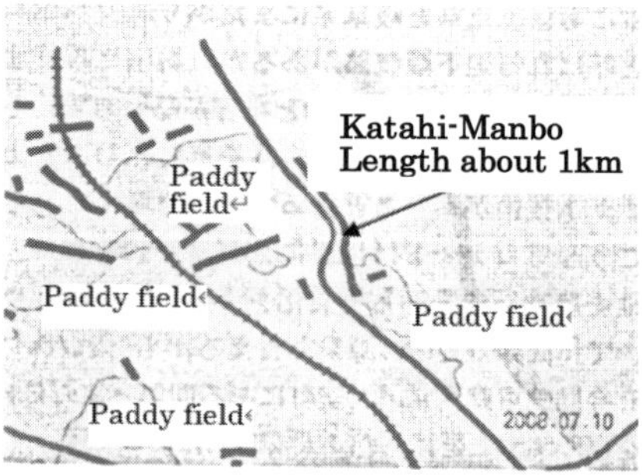

Fig.1.4 Neighborhood map of Katahi-Manbos (12)

Inabe City, Mie Prefecture, and these are located 2-3m underground from the water source at the foot of the mountain in order to solve the chronic water shortage in the late Edo period (constructed in about in 1770). A large number of Manbos (about 300) were made by excavating horizontal holes to guide water.

Photo 1.3 Katase-Mambo, Outlet of side pit (12)

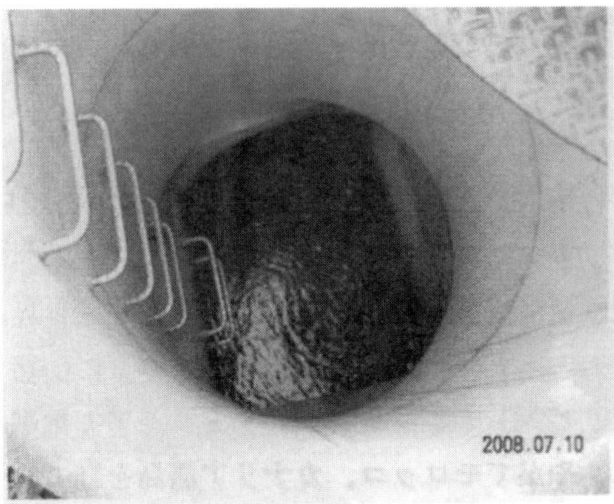

Photo 1.4 Katase-Mambo vertical

Photo 1.3 is the water outlet of "Katahi-Manbo" measuring about 1 km in length, and Photo 1.4 is the vertical shaft currently hardened with concrete. In this way, the qanat has spread

all over the world and has greatly nurtured agriculture, industry and culture in each
land, so it may be said that it is the source of the world's culture.
This is just a good example about the Silk Road of Flow Technology.

1.1.3 Greek-Roman Aqueduct (2)(5)

As mentioned above, the qanats were spread to the Greek-Roman world, but the second type of waterway in ancient Greece is the oldest. Stone-supported pipeline laid from the water source through a tunnel for the shortest route. An example is the Samos Aqueduct, which seems to date from the 6th century BC. Herodotus regarded this as one of the three major Greek works. The length of the Samos tunnel was about 1000m or more, and it was said that it had to be bent sharply at the end because it digging started from both sides and there was about a 5m step at the jointing point.

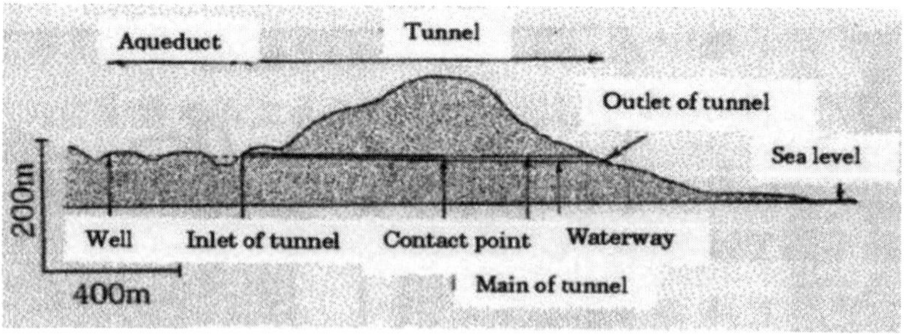

Fig. 1.5 Cross section of Samos Aqueduct

Fig.1.5 shows the Samos Aqueduct.

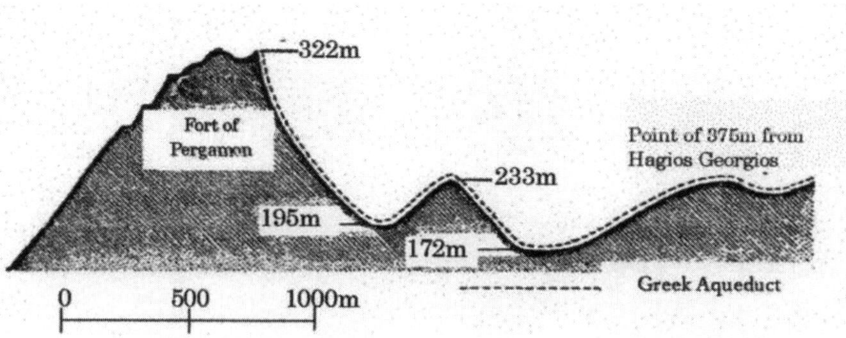

Fig. 1.6 Pergamon Aqueduct [1]

In Asia Minor, the Greek introduced a bold new hydropower device, called a siphon. It uses the gravitational energy of the water source to avoid costly tunnel construction.

The Pergamon Aqueduct built around 180 BC is a classic example of such siphon use. (See Fig. 1.6)

Water was supplied from a mountainous water source (1174m high) to two water tanks at Agios Georgios (375m high) in the east of city, using the water head. From there, it was led to a fort (332m in height) over two valleys (172m and 195m in height) and a peak sandwiched between them. Therefore, in the lower valley, the pressure of 20 atm would

Fig.1.7 Greek women taking a shower (From the back painting pot [1]

be applied due to the head difference, so it is thought that metal tubes were used.

Fig.1.7 is a picture of Greek women taking showers at the end of these channels. It's just amazing that there was a shower at around the 6th century BC!!

The Roman royal government is said to have continued from 753 B.C.E. to 510 B.C.E, but around 520 BC, at the end of the royal government, a sewer called "Cloaca Maxima" was constructed.

Fig.1.8 shows "Map of central Rome during the Roman Empire". Cloaca Maxima is indicated by a thick line.

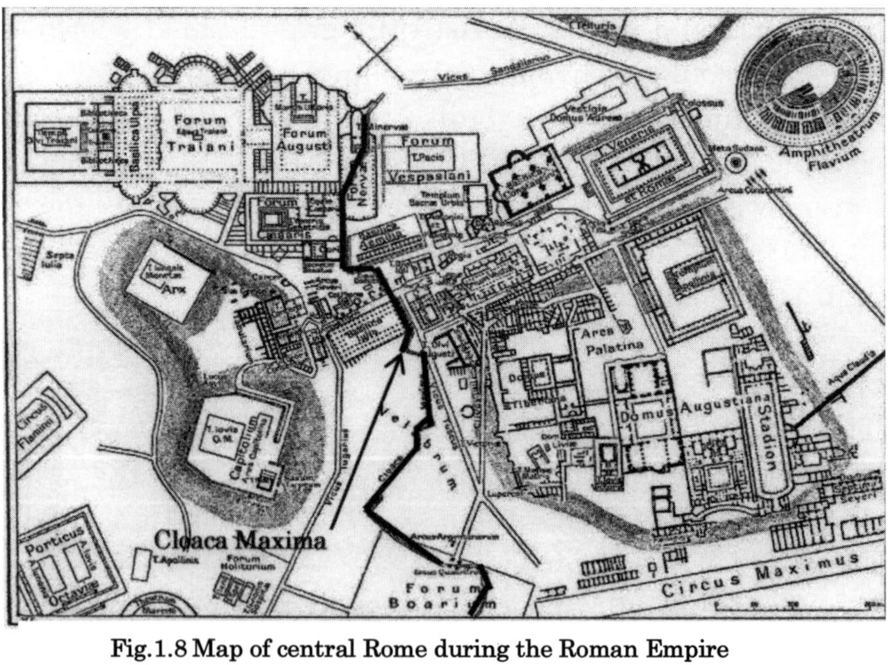

Fig.1.8 Map of central Rome during the Roman Empire by Yahoo USA Wikipedia (1

Its name means "maximum sewage". According to folklore, it was built by Tarquinius Priscus (616~578B.C.), king of the Roman kingdom, around 600 BC. (Source: Free encyclopedia Wikipedia Japan) Photo 1.5 shows Drainage outlet of Cloaca recently.

Photo 1.5 Drainage outlet of Cloaca Maxima, recently (13)

However, about 200 years passed before the birth of the water supply.

Roman Aqueducts are widely known for magnificent arches. Between 100 years from the middle of the 3rd century BC, when Rome had completely unified the Italian Peninsula, four waterworks, including the "Appian Waterworks", of the world's first real water supply were realized. The Appian Aqueduct was built around 300 BC by a blind general inspector and engineer called Appius Claudius.

Furthermore, in the early 1st century, when "Peace in Rome" was pronounced, Rome's population had grown to 500,000, and a total of 11 waterworks had been completed, including three from this era and others from previous era.

Fig. 1.9 shows the route map of Roman aqueduct bridges in Rome today.

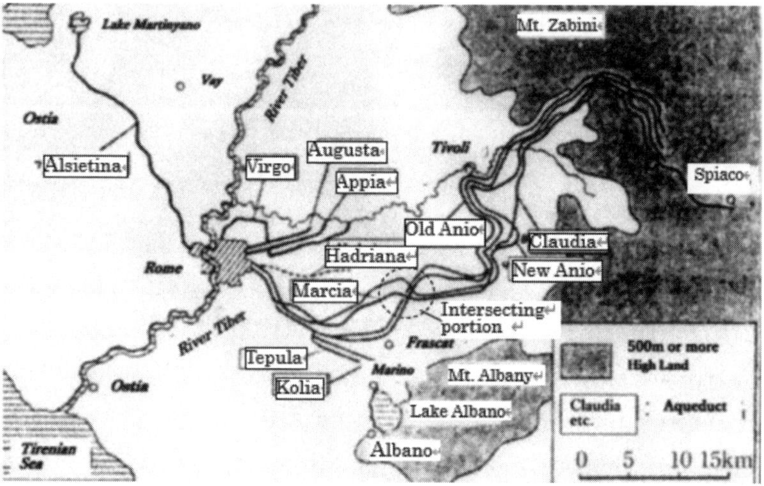

Fig.1.9 Route map of Roman Aqueducts today (1)

Photo1.4 shows the Ruin of New Anio Aqueduct and Photo 1.5 also shows the current site where three aqueducts of

Photo.1.4 Ruins of New Anio Aqueduct

Photo.1.5 Ruins of intersecting Roman aqueducts

Marcia, Claudia and New Anio intersect. As shown in Fig.1.10,
the water from these Aqueducts supplied three uses (public bathes, public fountains, private housings)

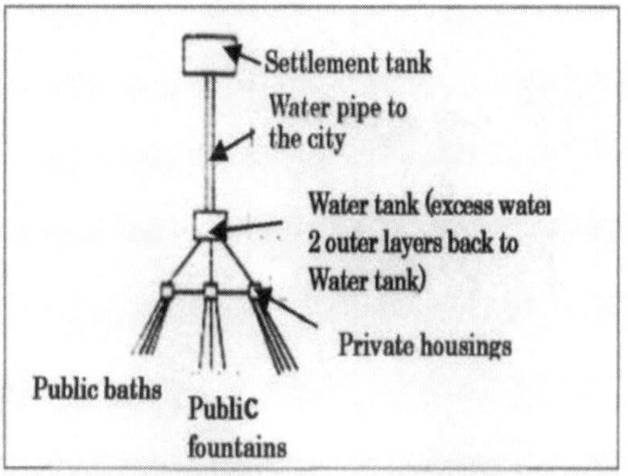

Fig.1.10 Roman water supply system (4)

How much water did the Roman aqueduct supply?
In this regard, scientist Heron of this era proposed a precise measurement method.
Until then, the flowrate was determined by the cross-sectional area of the flow, but Heron first pointed out that the velocity
of the flow was related to it, and adopted this as a measurement factor.
It is not clear how the flow velocity was measured, but the amount of water sent by the Roman aqueduct measured by Heron, was said to be 990,000 m^3 / day. Assuming that the population of Rome at that time was 1 million, the amount of water sent per person would be about 1,000 liters. Comparing this with the amount of water delivered per person in modern times, it seems to be too much, but it is

presumed that the Roman aqueducts were always allowing for the water continuously running.

1.1.4 Universal Engineer Heron

(3)

Portrait of Heron

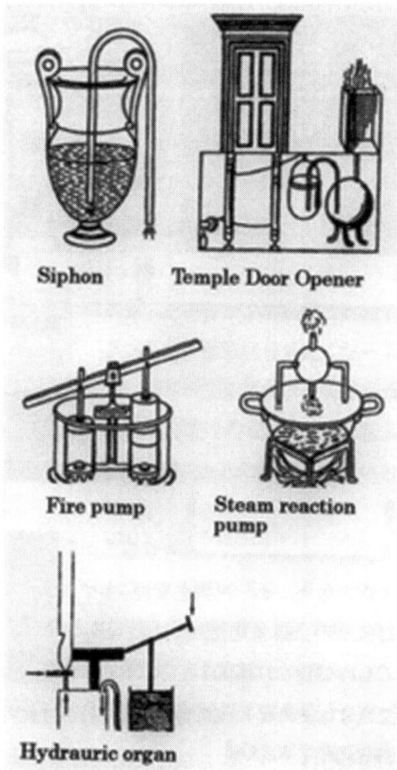

Fig.1.11 Fluidic machines were Invented by Heron

According to Dannemann's "History of Natural Sciences", Fig,1,11 shows Heron's inventions which utilized the properties of various fluids.

The most famous of these figures is the Temple door opener, which will be discussed later. Also, the principle of siphon was applied to the above-mentioned waterways of Pergamon.

As mentioned earlier, it seems that Heron was the first to propose the concept of flowrate, and Kesakatu Koizumi said

“Among the many Greek engineers, no one would have as good development ability as Heron or created a wide variety of instruments and devices.”

If people who are comparable to Heron in the same era and pre-modern times are put into a list, it would include Leonardo da Vinci and Rohan who invented the scale (steel square type) for carpenter in China”. Much of Heron's original writings and designs are lost, but some of his works were preserved, including manuscripts from the Eastern Roman Empire and in Latin or Arabic translations.

Heron was also a genius in mechanical technology, but the main feature of many of the instruments and devices he developed would be that he knew and used the properties of gas and hot air.

1.2 Ancient Fluidic Technology

1.2.1 Appearance of the Water Wheel

In ancient times, humans were the only driving force until the appearance of water wheels, and it seems that huge ruins such as the pyramids were also built by human power. Water wheels appeared around the 1st century BC. A person named Antipater of Thessalonica in Greek wrote the following poem about the emergence of a water mill. “Stop pulling, women working in water wheels. Even if a chicken crows and announces at the break of dawn, sleep slowly, because goddess Demeter ordered nymphs to do your handiwork. Nymphs jump down to the top of wheels and turn mandrels, and they turn heavy, hollowed stone mills with the rotating spokes"

The momentum of squirting water is anthropomorphized as The nymphs in this poem.

Thus, the oldest known water wheel was to mill grain. The most primitive mill is a Greek one called the Norwegian type water mill. Fig. 1.12 shows the principle, in which (a) is a "horizontal" water wheel with blades that rotate horizontally, with the upper end of shaft fixed to the upper millstone. The small amount of torrent hits blades from the oblique ditch. (b) is a "vertical" water wheel where the blades rotating vertically. This water wheel going back to the 1st century BC was the origin of the subsequent Pelton and Francis turbines, and it

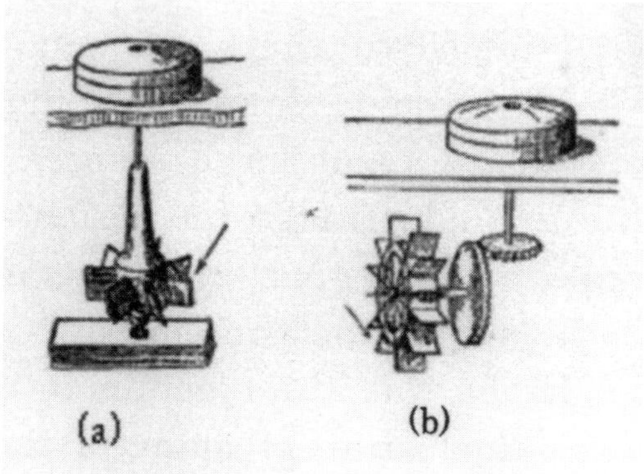

Fig.1.12 Norwegian water wheel (2)

Photo 1.6 Water wheel remaining

Photo 1.7 Wooden gear that rotates

is also the origin of the principle of the turbine flow meter.
I traveled in the past to the town of Ruurlo near the German border in the Netherlands for my work and was able to see the actual milling water-wheel at a restaurant where I was invited for dinner. The scene is shown in Photo 1.6, which has the shape shown in (b) in Fig. 1.21 and the stone mill that was rotated by a splendid transmission device (wooden gear) in the basement as shown in Photo 1.7.
Norwegian water-wheels became widespread to Ireland and China in the 3rd and 4th centuries. These were well known and built in Denmark until the beginning of the era, they spread rapidly in the centuries that followed, and increased significantly in the late Middle Ages.

1.2.2 Pythagorean Scholars (2)(8)(9)

The Pythagorean scholars who succeeded Pythagoras were good at making mechanical devices as well as mathematics.
Among them, it is said that Archimedes (BC 287 ~ 212) of Syracuse was the greatest person of all mechanical scholars and also one of the greatest mathematicians. From the perspective of fluid dynamics, he first discovered the "principle of buoyancy." The famous following episode remains. Archimedes was ordered by King Hieron I of

Photo 1.8 Archimedes

Syracuse to check if a golden crown made by an alchemist, it was not mixed with silver. He was thinking about this problem while entering a public bath one day. Perhaps he was inspired by the overflow of hot water when he took a bath. He found a solution and screamed "I found it!" and jumped out of the bathhouse and ran naked through the city.

The solution was to put a lump of pure gold and a lump of sterling silver, which weigh the same as the crown, in a container full of water, measured the amount of water that overflows, and compare it with the amount of water that overflows when the crown was soaked. The density of pure gold is 19.3 g / cm^3, while the density of sterling silver is 10.5 g / cm^3, so a gold crown mixed with silver should overflow more than a pure gold crown.

If the amount of water that overflows when a crown was soaked was more than when the mass of pure gold, the crown was mixed with silver. What was the fate of the alchemist due to this solution?

Archimedes was also familiar with the principle of leverage, and when the Roman army invaded Syracuse, he destroyed the enemy ships with a large catapult or lifted the enemy ship with a large iron claw. As a result, the Roman army suffered a great deal.

When immersing himself in research while looking at geometric figures on the street, a Roman soldier approached him with a sword. He was sadly stabbed to death by the soldier because he said, "You! Keep away from my figures."

It is said that in mourning this, the Roman General Marcellus who acknowledged his many scientific achievements and talents, was so sad that he built a tomb, which included Archimedes' proposition "The volume of sphere is equal 2/3 of the one of circumscribed cylinder." (See Fig. 1.13)

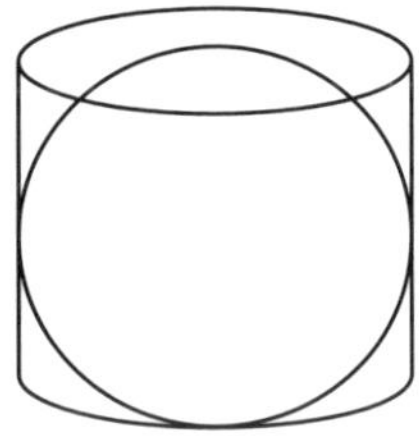

Fig1.13 Figure of Archimedes' proposition [2]

Archimedes and the successors of Archytas who believed to be the inventor of screw that appeared especially in Alexandria.

It seems that Ctesibius discovered the elasticity of air and invented weapons that used compressed air, suction pumps and push-up pumps. The successor to Ctesibius was Heron, mentioned in the previous section. Heron used the elasticity of air to create an interesting device shown in Fig. 1.14. To explain this principle. When the fire is lit on an altar (1), air inside the altar is heated and expands into the container at the bottom. The air flows and pushes the water inside of container (2), water flows into the next bucket (3), and then the bucket is lowered by the increase of weight to rotate two

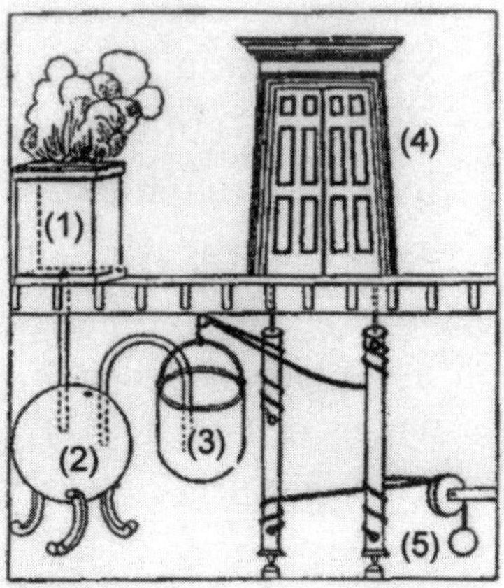

Fig.1.14 Heron's temple door switchgear [2]

steel shafts and opens a door (4). The door opens to left and right. Then the balance weight (5) at the right end rises. Perhaps this gimmick below was hidden, so people who saw this device for the first time would have been surprised and thought that the automatically opening door was God's work or magic. This is a respectable fluid machine.

1.2.3 Water Lifting Device

By the way, in the previous section (section 1.1.2), the development history of irrigation and water supply were explained, but these require water-lifting technology.

Advances in mechanical lifting devices from the simple discontinuous water lifting device such as splashing had been achieved by the application of the two Greek inventions of toothed wheels and screws.

Fig. 1.15 Recent "Saqiya" turned by a camel (2)

The invention of the toothed wheel is also said to be by Archimedes.

The "Persian water wheel" known as "Saqiya" is a vertical water wheel with pots that is rotated, by hanging this gear on a horizontal water wheel that humans or animals turn.

Fig.1.15 is a recent "Saqiya" (pots-wheel) that is turned by a camel to raise the water of the Nile River, but it seems that an ancient one still remains as it is because the structure is extremely rough.

Fig.1.16 shows the three types of water lifting devices classified by Vitruvius around 50 AD. (a) is a pulley bucket method, (b) & (c) are "Scooping wheels", and (d) & (e) are "Circle drum wheel" divided into eight. These are rotated by the flow of river. In Egypt, spinning gimmicks gradually entered the Hellenistic period. "Saqiya" were first described in the 2nd century BC.

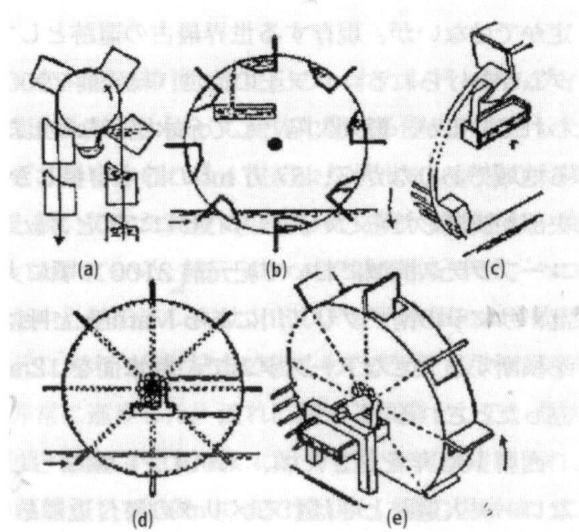

Fig.1.16 Three types of pumping machines

Up to the Roman era, in Egypt, "Circle drum wheels" had begun to replace the "Saqiya" and water was abundantly supplied by turning them by foot to irrigate vegetable gardens and to dissolve salt in salt farms. The Saqiya is still used in India, Egypt and other parts of the Middle East, and in the Iberian Peninsula and the Balearic Islands. In modern classification, lifting water lifting devices powered by animals are called Saqiya, and those that use the flow of river water as in Fig.1.16 (d) and (e) are classified as Noria.

As an example of famous Noria, Hummer's Noria is a number of Noria (wheels of pots) along the Orontes River in Hummer, Syria. Only 17 of the original Noria remain. These are currently largely unused and are for aesthetic purposes. These are known as "the most spectacular Noria ever built". Hummer Noria was submitted by the Arab Republic of Syria in June 1999 as a temporary World Heritage site. (See Photo 1.10)

Photo 1.10 Hummer Noria

The invention of the screw is attributed to Archytas (c. 400 BC), but it was the Archimedes irrigation screw from the earliest clear report. According to Diorodos (1st century BC), it was said that this snake like machine irrigated the Nile delta. Figure 1.17 is an Archimedes'

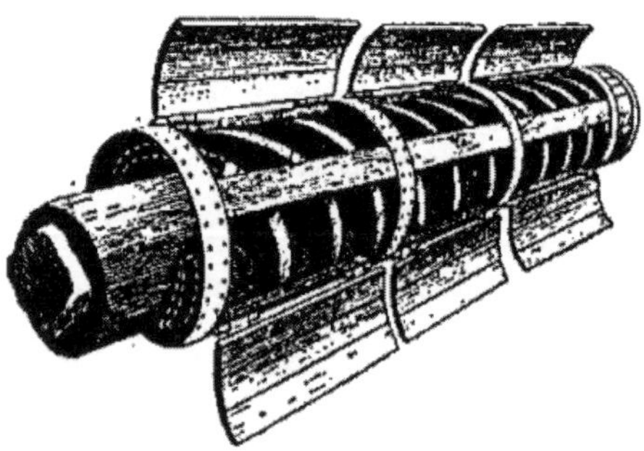

Fig. 1.17 Archimedes' screw [2]

screw made from oak for lifting water excavated from a mine in Spain. The diameter of shaft is 61 cm, and spiral thread is about 5 cm. Archimedes is famous for discovering the "principle of buoyancy", and can be said to be the originator of the invention of fluid machine.

1.2.4. History of Dams

The Ancients devised dams as reservoirs for water supply. The birthplace of the ancient dam is uncertain, but oldest surviving archaeological site in the world is Sadd el-kafara dam in Egypt. It is said that to have been built during the reign of King Khufu (2900-2877 BC), but the biggest defect was that there was no spillway. Although it was in an area that received torrential rain, the dam had a storage capacity of only 570,000 m^3, therefore it seems that the central part of the dam collapsed shortly after its completion.

On the other hand, irrigation in Babylonia and Assyria made huge progress in the Tigris and Euphrates basin around 2100 BC. An old dam called Marduk, located on the South Tigris River in Samara, north of Baghdad, was a large earth dam that crosses the Tigris River and is said to have raised the river surface by as much as 12m, helping to contain a large amount of water. Around 100 A.D. a wonderful dam appeared in Spain. It is the Proserpina dam with a height of 19m and a length of 427m near the town of Merida, which was called as a huge Roman ruin.

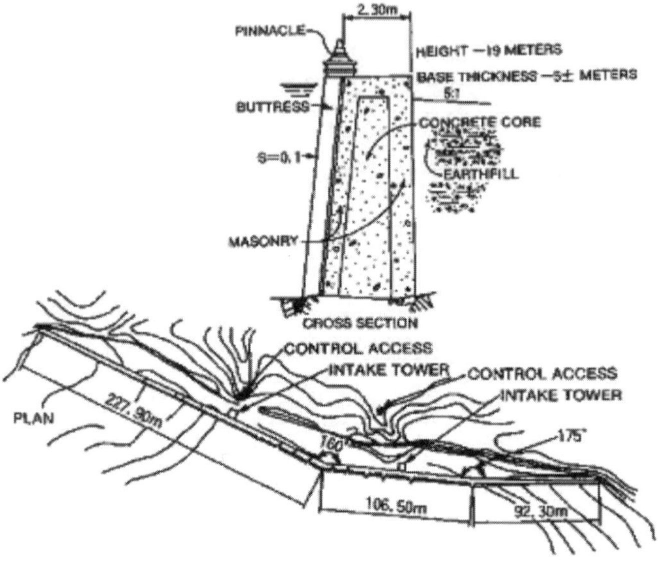

Fig.1.18 Drawing of Proserpina Dam (7)

This is shown in Fig.1.18. It is the Roman gravity dam in Badajoz (province), Extremadura, Spain, dating from the 1st or 2nd century AD. This is, the largest dam built by the Romans and the oldest dam in use today. The Cormalbo Dam was also one of those built around this time.

Photo of Proserpina Dam

1.2.5 Development of Chinese Science

Now, the science that blossomed in the ancient Orient did not change much until the appearance of Leonardo da Vinci in the 15th century. The era without historical discoveries and inventions about fluids had finished. It is said that this was because there was a great migration of the Germanic people from the 4th century, the highly developed civilization that continued in Greece and Rome had been destroyed, and valuable literature was lost.

In Europe in the middle of the 7th century, Alexandria which was a center where ancient engineers such as Heron were active had been fallen in the Crusades. From this time Europe entered the dark of the Middle Ages, which continued until the 15th century. On the contrary, in Eastern China, weights and measures had progressed a lot during these ages. As a business related to flow, the Grand Canals had been built from the Spring and Autumn period to the Kan dynasty (around the 8th century BC to the 3rd century AD), the length of the canals reached 2400 km in the Sui dynasty (6th century AD). Emperor Yang Jian during the Sui dynasty replaced these waterways, expanded and rearranged the river channel as a whole.

Finally, he made a main artery in order to connect the center to the breadbasket of Gangnam, such as a ditch that connected the Yellow River and Huai River through the north and south, and another ditch that reached the outskirts of Beijing from the Yellow River etc.

Furthermore, when the capital was moved to Beijing in the period from the Yuan (13th and 14th centuries) to the Ming (14th to 17th centuries), a new waterway was created that crossed the north and south in a straight line. This is the world's longest grand canal of 1800km still in use now. In fluid instruments, "bellows" was used as a metallurgical technique, and in the Han dynasty (up to the 3rd century), box-shaped ones that blows air by moving the piston back and forth, known as "Hai", began to be used and has remained in use until recently in China and Japan. In addition, the type of "Water-Hai" that can be moved by human power and other methods that can blow a large amount of air via hydraulic power had been also developed. A large amount of cast iron refining in the Han dynasty was made possible by this hydraulic method. In Japan, where paddy rice had been the main food, technology for water was developed at an early stage. In Japan ancient books "Fudoki" and "Kojiki", something similar to a siphon called "Takahi" appeared.

So far, the history of ancient technologies has been described, especially those related to fluid. From the next chapter, the contents will move to the history of flow measurement technology from the Middle Ages to the modern era.

Reference

(1) Charles Singer "A History of Technology (2)" Tikuma, 1963

(2) Charles Singer "A History of Technology (4)" Tikuma, 1963

(3) Kesakatu Koizumi "Ancient Fluid Technology" Oval News No.281,1986

(4) "Fluid culture" Oval SPECIALTY No.6
(5) "Fluid culture" Oval SPECIALTY No.8
(6) Kesakatu Koizumi "Great oriental technology" Oval SPECIALTY No.2 1988
(7) "Fluid culture" Oval SPECIALTY No.9
(8) Katuhiko Asakura "World Human Anecdote Dictionary", Kadokawa
(9) Yutaka Hosoi "Fluid Dynamics (1)" Tokyo Denki University Press Div.
(10) Inter net YAHOO JAPAN "Egypt travelogue"
(11) Noboru Koyano "Appear of ancient Egyptian city civilization" Kokin Book shop, 1998
(12) Yutaka Ogawa "The history of Japanese water flow technologies" Bulletin of Society of Historical Metrology, Japan Vol.31 No.1 (No.36) 2009
(13) Source: Free encyclopedia Wikipedia Japan, 2021
(14) Source: Cloaca Maxima in Rome during Roman Empire, Yahoo USA Wikipedia

2. Medieval Science Technology

2.1Flowering of the Scientific Renaissance

2.1.1 From the Dark Ages to the Renaissance (2)

The plunder of Rome city by Visigoth (King Alaric) for three days from August 24, 410 BC was the first step in the fall of ancient culture that Roman citizens saw themselves.

Following the destruction of Western Roman Empire by the Germanic mercenary chief Odoacer, the closure of Academia in Athens by the Justinian Empire of Eastern Roman (Byzantine) in 529, and the fall of Alexandria by Saracens in 641, the bases of ancient culture had disappeared in flame and smoke and the Dark Ages began. However, it wasn't if previous culture was completely broken and suddenly blossomed in the Renaissance. It is said that the knowledge of ancient culture was passed on to modern culture, by the monastery system of the Benedictine order, which was the central force in the new European order. Byzantine culture flowed through Constantinople and Saracen culture via Arabia.

In particular, it is said that Arabian science that had been constructed by the Saracens, “not only unknowingly held and defended the huge body of ancient science that became a [refrigerator] in dark medieval period”, but also formed the mainstream of world science history by adding creative flows in the sections of mathematics, astronomy, and chemistry, etc.

Since chemical terms such as alkali and alcohol are Arabic, we can see a part of this.

Photo 2.1 shows a Saracen celestial navigator as an example of astronomy. In Europe, there was Scholastic philosophy, centered in Italy from the 9th Century to 13th Century. It was a discipline that demonstrated and defended how the church's beliefs made sense, and since it was sourced from Greek classics, "the Restoration of Greek classics" was the most important task. An Arabian followed a roundabout way of translating

Photo 2.1 celestial navigator of Saracen (2)

Greek classics from Greek to Hebrew, or to Arabic through ancient Greek, and finally to Latin, something like that.

From the 11th Century to 13th Century, the "Great Expedition of Crusaders" had been conducted. They had discovered the Byzantine and Saracen cultures mentioned earlier, and a great exchange of the three major civilizations of Greek, Rome, and Arabia had taken place here. The decline of religious scholarship and the fall of the feudal system due to the failure of Crusades caused the collapse of scholastic worldview. As a result, the era shifted to the Italian Renaissance, which began in the 13th century.

Concerning fluid machinery in this dark age, only the appearance of windmills in France in the early 12th century can be mentioned.

2.1.2 Significance of the Renaissance (2)

Let's review a little about the Renaissance era.

A Medieval German saying says, "Farmers are no different from bulls, but they just have no horns". From the huge feudal medieval society, where such oppressed serf-like peasants accounted for about 90%, they discovered new human beings and the world, and released from the bondage of God. As a result, human freedom and rights were enjoyed in the early modern period. This transition of society was the Renaissance that meant human regeneration.

The first stage of the Renaissance was in international cities such as Florence and Venice in Italy from 14th to 16th centuries. It is said that the forerunner was opened by two poems, the "Divine Comedy" by Dante around 1318 and "Decameron" by Boccaccio in 1352. The reconstruction was followed by paintings such as "Mona Lisa" and "The Last Supper" by Leonardo da Vinci and "Madonna" by Rafael, and sculptures such as "David" and "Moses" by Michelangelo.

Looking at the Renaissance period from the aspect of science and technology, its curtain was opened by Copernicus (Heliocentrism) and Vesalius (structure of the human body), and it was completed by Galilei and Newton, but concerning experimental methods and the spirit of modern natural science, pioneers were the English monk Roger Bacon in 13th-Century and Leonardo da Vinci, the all-purpose genius in 15th-Century.

Roger Bacon (UK, 1214-1285) was the first scientist in the modern sense, because he was active in a wide range of fields such as optics, astronomy, geography, mechanics, chemistry, mathematics, based on observation and experimentation.

In his book "Ops Majus", he says: "There are two modes of knowledge: through argument and through experience. Argument brings conclusions and compels us to concede them, but it does not cause certainty nor remove doubts that the mind may rest in truth, unless this is provided by experience."

Photo 2.2 Portrait of Roger Bacon

He spent 20 years in prison before and after, because he respected his experience contrary to the scholastic style of speculation and made unforgivable criticism of the authorities, he finally became first person on the list of victims of religious persecution.

The three major inventions of ocean navigation, gunpowder and printing are representative of the technology of the Renaissance period. Owing to advances in ocean navigation, Columbus discovered the New World in 1492. Gunpowder was just a big signal of a major shift from the Middle Ages to the early modern period. In fact, the fall of the Byzantine Empire (1453) was due to an overwhelming attack by the bronze cannon of the emerging Turkish army Mohammed II. The invention of the printing press by Gutenberg eliminated the lack of preservation and transmission of medieval knowledge, and increased the intellectual vitality in the world.

2.1.3 Leonardo da Vinci [1][2][3][4]

The renaissance of science began in the middle of the 16th century and bloomed in the 17th. Few remarkable scientists were found in the first half of the 16th century, but the only thing that must be touched upon when considering the history of science and technology is Leonardo da Vinci (Italy, 1452-1519) who painted the famous smiling beauty "Mona Lisa" (See Photo 2.2). He was an artist, philosopher, engineer, and scientist of noble thought, bold speculation, and brilliant practice. He wrote a marvelous achievement in his thousands of pages of notebooks, which were discovered at the end of 18th century, 250 years after his death. For this reason, it made little contribution to scientific progress. The reason why he is also a pioneer in modern technology and science is that the words "Try this" appears many times at the ends of notebooks.

Photo 2.3 Mona Lisa in the Louvre Museum [4]

Respect for experimentation and experience is also clearly demonstrated in the following words:

"Experiment is the interpreter of nature. Experiments never deceive. It is our judgment which sometimes deceives itself because it expects results which experiment refuses. We must consult experiment, varying the circumstances, until we have deduced general rules, for experiment alone can furnish reliable rules."

It would be a clear statement to suggest the attitude that should be taken as an engineer and a scientist.

In Florence in the 15th century, many painters had a keen interest in landscapes, flowers and trees. Leonardo da Vinci was also influenced by these, but since his observing power was more scientific than anyone else in the past, he left many sketches in his notebook that also show amazing vitality. For example, a huge notebook about water was drawn around 1490.

Many drawings of swirling water are left, and in Fig. 2.1,

Fig.2.1 Drawing of Swirling water (1)

Fig.2.2 Water study note: Profile of an old man sitting on a ledge and facing right (1)

and 2.2 there is a study of swirling water released from an object that blocks the flow, and below that, it symbolizes the phenomenon of water entrainment at the falling point of the discharged water as drawn like a whirlpool flower pattern. This should be seen as a skill that can be achieved through his excellent insight as a scientist and his exceptional creativity as an artist.

As the vortex from an object that blocks flow was later analytically elucidated by Karman 1911, it is called the Karman vortex, but Leonardo left many of these vortex phenomena. His study of water is also an idea associated with the movement of blood in human anatomy Therefore, it is natural that a human figure and the whirlpool of water appears side by side like in this drawing. It is surprising that the depiction like Karman vortex in Fig.2.2 is more elaborate than in Fig. 2.1. The figure of an old man is often regarded as an idealized self-portrait.

As for why he leaves people and physical phenomena side by side in this way, Carlo Pedretti said, "Leonardo thought of the earth as an organism that resembled the human body. His study of water is also an idea associated with the movement of blood in human anatomy. Therefore, it is natural that a human figure and the whirlpool of water appear side by side like this drawing."

Fig.2.3 With the study of water Blocked by single board of impediment and Note (1)

Finally, Fig 2.3 is the most interesting one. In addition to the vortex phenomenon emitted from the obstacle, the shape of various obstacles (bluff body) is written just below, and there is something like a streamline in the lower row. What is written in the explanation of this note? Concerning the movement of water, he who focused on experimentation and observation, conducted numerous observed experiments such as vortices, falling water, and open channel flow, and had left an observation note suggesting what became known as "The principle of fluid continuity" that is the conservation of mass and "Pascal's principle" that is the principle of hydraulic press force. Further, he left a sketch of streamline gradually tapering in order to reduce air resistance. He also dreamed of flying, and devised helicopter and ornithopter with a spiral wing. Hygrometer and anemometer are also considered his inventions.

As mentioned above, he was a pioneer of modern technology and science, but from his keen observation of the water vortex, he is naturally a pioneer of fluid dynamics and is named alongside Archimedes and Heron.

Photo 2.4 Galileo Galilei [(3)]

2.1.4 Flowering of the Science Renaissance

In the latter half of the 16th century, brilliant people who decorated the beginning of the story of early modern scientists such as Copernicus, Vesalius, Tycho Brahe,

Galileo, Kepler and Gilbert appeared. As you know, Galileo Galilei (Italy, 1564-1642), another genius who represents the Italian Renaissance era along with Da Vinci, thinks of his kinematics as celestial movement rather than fluid movement.

y Although there are few results of research on fluids, he stated that pumps cannot suck up when the suction height is about 9 m or more. Since Aristoteles stated that "nature hates vacuum", everyone had implicitly affirmed the existence of vacuum which was negative.

Now, in this way, time flows from the dawn of the scientific renaissance to the heyday like hundred flowers blooming in profusion.

Moving on to Japan, it was just the darkest ages of the Middle Era.

In the 7th century, in 646, the "Taika Reform" was carried out by Prince Nakano Oue, which produced "Opening Japan". At this time, "old affairs "Kojiki (literally, Old history)", "Fudoki (literally, essay)", and "Manyousyu (literally, oldest poetry collection)" had been compiled, and in the 11th century, excellent literary works such as "The Genji Story" by Murasaki Shikibu and "Houjouki" by Kamo no Cyoumei in the 13th century appeared. From the beginning of the 16th century, when Leonardo da Vinci was active, Japan entered the Warring States period and changed into a world of constant war.

If one theme "flow" is chosen, there is a famous sentence in Houjouki that Kamo no Cyoumei compered the impermanence of world to the flow of river in 1212. “The flow of the river is ceaseless; and its water is never the same. The foam that floats in the pools. Now gathering, now vanishing. Never lasts long. So, it is with man and all his dwelling places on this earth.”

How emotional this is when compared to the scientific observations of the Leonardo da Vinci.

2.2. Sprouting of Measurement Technology

2.2.1 The Era of Genius (2)

In the previous section (Clause 2.1.3), Leonardo da Vinci was described, as a pioneer of modern technology, here is a little more detail.

Now since there is a portrait of him when he was young, shown in Photo 2.7, but it is unknown whether it was from that era or later. It is the Da Vinci wearing a beret-style hat. What a dandy! Now, the era shifts to the 17th century,

Photo 2.7
Young Leonardo Da Vinci (2)

Mitsutomo Yuasa said "This era was a brilliant century in which modern science was completely established, but in the history of technology, it was a transitional era leading to the Industrial Revolution, and no creative invention was seen. It was all about the advancement and improvement of experimental machinery and equipment that natural science research needed, but this century was thought to be the era when many measurement technologies have been born."

This era is thought to be a time when measurement technology had mainly grown up. In the first half of this century, brilliant members well known to posterity appeared such as Galilei (telescope, astronomical dialogue, dynamic dialogue), Kepler (planetary motion), Descartes (analytic geometry, kinematics), Harvey (blood circulation), Torricelli (vacuum), Pascal (atmospheric pressure), and it was exactly the "era of genius". In these, the following sections I focus on Torricelli and Pascal, who had a significant influence on the history of fluid measurement, because their researches on pressure are an essential state quantity for flow measurement.

Photo 2.8
Torricelli's portrait (2)

2.2.2 Fluid Measurement (5)

Torricelli (Evangelista Torricelli, Italy, 1608-47) was a talented young Italian aristocrat. (Photo 2.8)

He studied under Galileo for a short time, and was particularly interested in Galileo's statement that "Pumps cannot be pumped up when the suction height is 9 m or higher."

On this issue, he began experimenting with the only metallic mercury, which is liquid at room temperature. Because mercury is 13 times heavier than water, he thought that the atmospheric pressure column would be only 1/13 height of water. Therefore, a glass tube with a length of 2 yards (about 1.8 m) was filled with mercury, and while pressing the entrance with a finger, it was immersed vertically in a dish filled with mercury. Then, the upper part of the glass tube became a space, and the height of the mercury column was about 30 inches (760 mm). When this operation is reproduced, it becomes as shown in Fig. 2.4. It is a historical moment for the invention of barometer. However, this barometer was not stable day by day. He said this was because the air pressure was changing and was measuring it

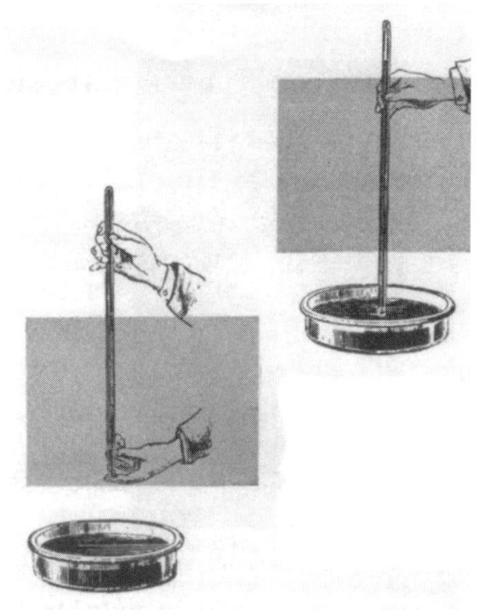

Fig.2.4 Atmospheric test of Torricelli (5)

accurately but he unfortunately died prematurely at the young age of 39 without proving this. It was Pascal (Blaise Pascal, France,1623-1662) who proved that a glass tube containing mercury could be a barometer. Pascal was very interested in Torricelli's experiments.

Photo 2.9 Pascal's portrait (2)

The time was September 1648. He was in Paris at that time, and he was guided by his brother-in-law, F. Perrier, to the Puy-de-Dom e Mountain at 1465 m above sea level in the Auvergne to conduct experiments. The brother-in-law placed barometers on several different slopes. If Torricelli's theory was correct, where the mountain was higher, the mercury column should be lower.

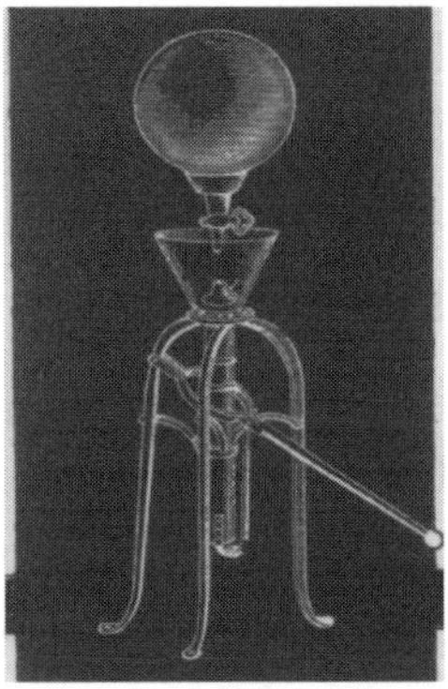

Fig.2.5 Air Pump (5)

The result exactly indicated that phenomena. At this time, Perrier said, "We were excited by surprise and joy."

Pascal also discovered the principle of Pascal that pressure is transmitted equally in all directions in stationary fluid.

Around the same time, in 1652, Otto von Guericke (Germany, 1602-1686) invented the air pump. (See Fig.2.5) He was a young man who lived in the midst of the horror of the Thirty Years War.

He was unemployed in 1631 and left his hometown of Magdeburg, but later became the mayor after returning home.

The first air pump filled a barrel with water and attempted to drain it with the pump, but failed because of a leak. Next, he tried to pump out air inside a copper spherical shell, but the ball was buckled by atmospheric pressure and broke. However, in the end, an air pump with a considerable degree of vacuum was completed.

The most famous his experiment is the "Hemisphere of Magdeburg". This is two hemispherical shells carefully combined into one and the air inside was evacuated to create vacuum. Next, eight horses pulled both ends of the hemispheres to either side, but both could not be pulled apart. (See Photo 2.10)

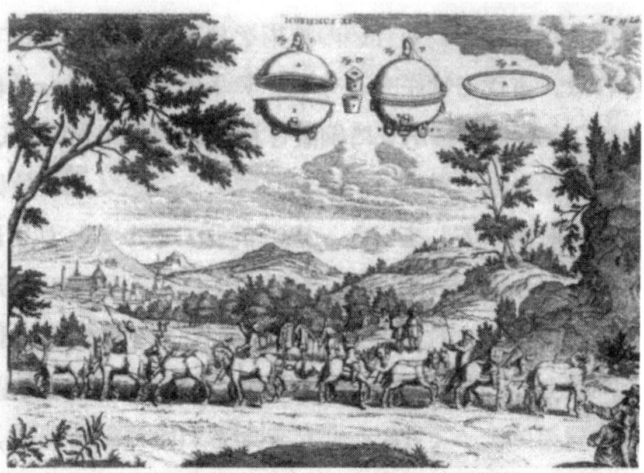

Photo 2.10 Experiment of "Hemisphere of Magdeburg" (5)

Was it an experiment to show how strong

the vacuum was? No, it was to show how strong the force was at atmospheric pressure.

After that, when air was put in, the two hemispheres were easily separated. Photo 2.10 shows the state of the experiment.

Another experimental Photo 2.11 shows a landscape painting in which an air pump made of a combination of a cylindrical case and a flat plate were evacuated and pulled apart by man power.

Boyle (Robert Boyle, UK 1627-1691) is classified as a

Photo 2.11 Experiment of air pressure (2)

Photo 2.12 Portrait of Boyle (18)

chemist, but he has a prominent result record for the so-called "Boyle-Charles' Law", especially for gaseous fluids. He discovered the elasticity of air using a U-shaped manometer with one end closed. When mercury was poured in the open end, the volume of air on the closed end side was visually compressed and measured.

This laid the foundation for the following famous formula: That is, when the pressure and temperature of the gas are represented by P and T, the volume is represented by V, and the gas constant is represented by R, the next equation consists.

$$PV = RT \text{------------------------------------}(1)$$

This is the origin of what was established.

2.2.3. Founding of the Academy

Nowadays, new science and technology researches travel around the world in a blink of an eye with the development of information equipment such as telephones, facsimiles, and E-mails etc. and the development of transportation such as aircraft etc. However, pre-modern researchers and engineers had to carry out their own researches and experiments. The establishment of academies were considered as organizations that shared each other's technologies.

In ancient, the academy in Athens of Greece is famous. However, Academia Secretorum Naturae (meaning a secret society of nature) which was founded in Naples in 1560, was the first one. It was suspected of Magic and "black technology" and was soon disbanded as a result.

Next, Accademia del Lincei (meaning the science of big wildcat) was founded in Rome in 1603 by Dela Porta. which is considered to be a respected and significant modern one due to its members which included Galileo.

Most importantly, Accademia del Cimento (Academy of Experiments) was founded in 1651 by two brothers.

The two brothers, who were studying under Galileo, invited scientists from home and abroad to make them using experimental equipment and to participate in the experiments.

Two of Galileo's disciples, Torricelli, and the prominent mathematician Vincenzo Viviani played a leading part in the spirit of this Invisible college.

Photo 2.13 shows the instruments used in the Accademia del Cimento.

Photo 2.13 Instruments used in Accademia Cimento (5)

Two things on the left are thermometers and the one on the right is a hydrometer (probably the current reference volume meter).

2.2.4. Significance of Newtonian Dynamics Formation

In the latter half of the 17th century, scientists who laid the foundations of modern science such as Boyle, Hook, Newton,

Photo 2.14 Portrait of Newton (2)

Leibniz and Huygens appeared one after another, but a biggest harvest in the 17th century would be the discovery of universal gravitation and the publication of "Principia" (1687).

It is also well known that the invention of the calculus method by Newton and Leibniz have been very useful as a means for solving scientific and technological problems since then.

Integral symbol written in Leibniz's letter in Fig. 2.6 is noteworthy.

The contents of "Dynamical Dialogue" of Galilei and "Principia" of Newton were very useful in this century.

But Torricelli and Pascal died early in life at the young age of 39. Regrettably. How much better inventions and discoveries they would have made if they had lived longer?

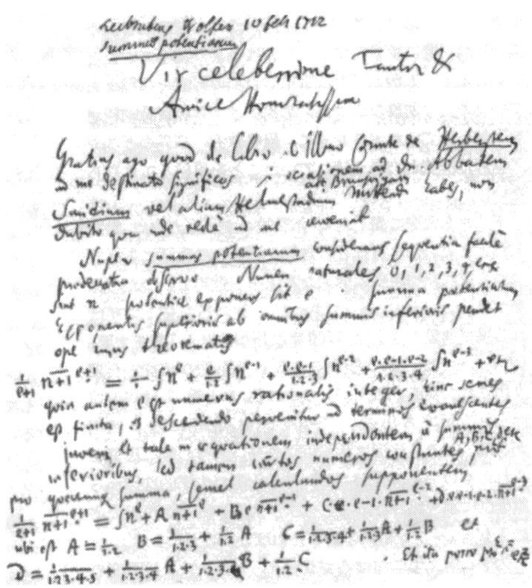

Fig.2.6 Letter from Leibniz [2]

2.3 Water-Wheel and Windmill Using Flow

Among things mentioned so far, only the measurement of atmospheric pressure had been done as a result, and it is still far from "flow measurement".

Therefore, the background history leading up to "flow measurement" will be described. So, the contents of this notation will be summarized with my interest first of all.

2.3.1 Medieval Water-Wheel (6)

In the Middle Ages, the application of hydropower to various processes of industry is said to have led to highly mechanized cotton spinning mills of the British Industrial Revolution.

From ancient times to early medieval times, the power of falling water was mostly used to mill wheat, but in the 16th century, in addition to water-wheels for metal melting, forging, cutting, rolling,

Fig.2.7 Cutlery manufacturing factory (Vegler, 1661) (6)

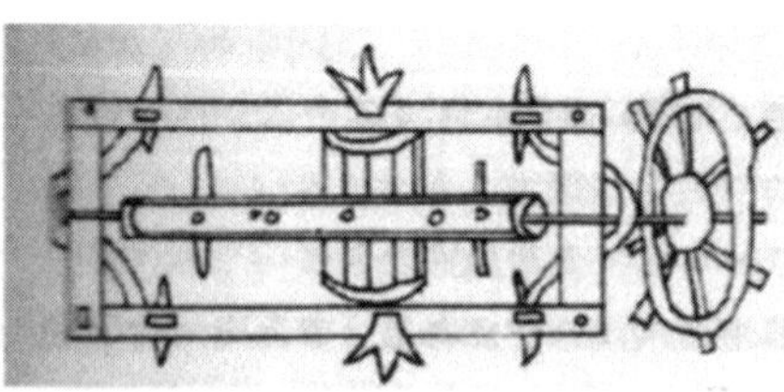

Fig.2.8 (a)Boring machine in 15th century (6)

Fig.2.8 (b) Boring machine in 17th century (6)

polishing, crushing, and punching had appeared. Fig.2.7 is a diagram of a water-wheel used in the knife manufacturing industry around 1660.

Photo 2.8 (a) is sketch of boring machine in 15th century and 2.8 (b) is an example of a water-wheel use for a boring machine in 17th century.

In East Germany, chain pumps with rags were used to drain tunnels. (See Photo 2.9)

Military demand contributed to the expansion of the use for hydropower in the metallurgical process. For example, for the first time in the period 1500-1750, water wheels were used as the power source for boring the barrels of musket

Photo 2.9 Chain pump with rags (6)

Photo.2.10 Boring and polishing machine of Gun (6)

guns or cannons (See Photo.2.10). In it, four horizontal drills are placed side by side and rotated by hydraulic power.

Also, there was a grinding wheel near the back wall, which was used to polish the outer surface of the barrel. In China as well, a water truck for pumping water to rice fields as shown in Fig. 2,9 was described in "Tiangong Kaiwu" published in 1637.

Fig.2.9 Chinese pumping wheel

2.3.2 Windmill History (7)(9)

The windmill is the root of inferential anemometer that is still used today. In Europe, old windmills can be seen everywhere. The windmill does not seem to have as clear history and geographical distribution as the water wheel.

In the early 10th century Persian manuscripts, there is a record that the horizontal windmill was first used in 644, but it is considered that it already existed in Egypt during the time of Moses.

From the 4th to 15th centuries, the Roman Empire to the Renaissance, advances in hydraulic machinery

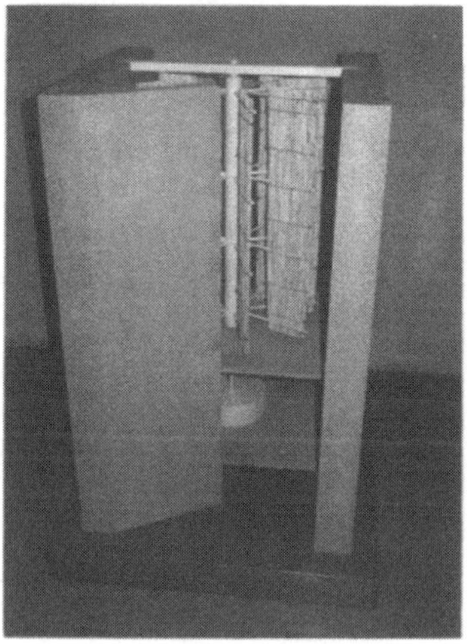

Photo 2.11 (a) Persian type of Windmill (7)

manifested itself in an increase in the quantity of water mills and windmills.

Especially until the 10th century, horizontal wind wheels were used for irrigation pumping in Persia.

Photo 2.11 (a) is the restored model.

Photo 2.11(b) shows vertical windmills in a town in northeastern Iran that has been in operation for 1000 years.

Photo 2.11(b) Vertical windmill in Iran

It is still unknown how this horizontal wheel was tied to the Mediterranean windmill, which seems to be the pioneer of today's vertical windmills. (See Photo 2.12)

It is said that there are still 6,000 pumping up wind wheels on Crete.

In fact, the vertical windmills used in Europe seem to have been a completely unique innovation, not originated from the Persian wind wheel, but might have evolved from the machine arranged from the water wheel for grain.

It was the Crusaders who spread the technology of this windmill and passed on the technology of

Photo 2.12 Sail-style pumping windmill on Crete (Greece) (7)

Arabian windmills to western Europe. On the other hand, due to the Islamic invasion of Mongolia, it was also conveyed to eastern China. Especially in the Song dynasty from the 10th century to the 13th century, windmill civilization expanded and a considerable number have been used in various places since then. There were the windmills in England shortly before 1,200 years.

However, since Pope Celestin III imposed a tithe tax on windmill, the numbers reduced. By the end of the 13th century, they became developed in Scandinavia.

The Netherlands has been below sea level since the beginning of the Middle Ages. The country was hit by continued floods, especially on November 18-19, 1421, with 72 villages destroyed. There, barriers were built to prevent seawater floods, and land was formed by draining the space between the coastal levees, but puddles and lakes still remained there. Pumping windmills were used for this solution.

Photo 2.13 Classic windmill in Zaanse Schans near Amsterdam (7)

Initially the direction of the blades of the windmill was fixed, but it has been improved so that the direction can be changed depending on the direction of wind. It is said that the number of Dutch windmills reached its peak around 1850 and there were 10,000 at that time.

Photo 2.13 shows three of the four classic Zaanse Schans windmills that everyone visits when traveling in the Netherlands. It is said that these have demonstrated milling, drainage, sawing, and mustard agitation.

2.3.3 History of Japanese Windmills and Water-Wheels

As represented by "When east wind is blowing, let plum give me it's smell, even though the owner gone, you don't forget spring" (1006). Songs and novels with the theme of the wind had been in Japan since ancient times. Japanese culture had been introduced from China first, from the Netherlands and Portugal in the Middle Ages next.

There is the fact that a considerable number of windmills were used in neighboring China, and it seems that the technology may be transmitted from the Netherlands and Portugal, which were major windmill countries. But there is no evidence of practical windmills used other than the "Kazaguruma" as a toy in Japan.

Since Japan has many mountains and heavy rain, it is blessed with water and waterwheels were developed. Even though the wind became the subject of the song, it seems that it was hated as a disaster such the typhoon. Also, even if a windmill was made, it would be destroyed by repeated typhoons, which was probably the main reason. The times will fly first, but the history of windmills in Japan seems to have started with the civilization of the Meiji era. Photo 2.14 is a picture of a pumping windmill at the Ferris Girls' School in Yokohama, which was known as the "Red Windmill School" in the middle of the Meiji era.

Photo 2.14 Red windmill at Ferris Girls' School (7)

So, what about the water wheels?

The history of Japanese water wheels is said to have begun on March 18 (610) during the Empress Suiko era according to "Nihon Shoki (the Chronicles of Japan)". The Goguryeo monk Doncyou introduced the Tengai (Water mortar). There is no ruin or painting in Japan, but in China, the Ming dynasty book "Tiangong Kaiwu" (17th century) likely has such a picture the shows this (Fig.2.10). Perhaps what had been in China for a long time would be transmitted to Goguryeo and introduced to Japan. While assimilated into the climate of Japan, it became widespread

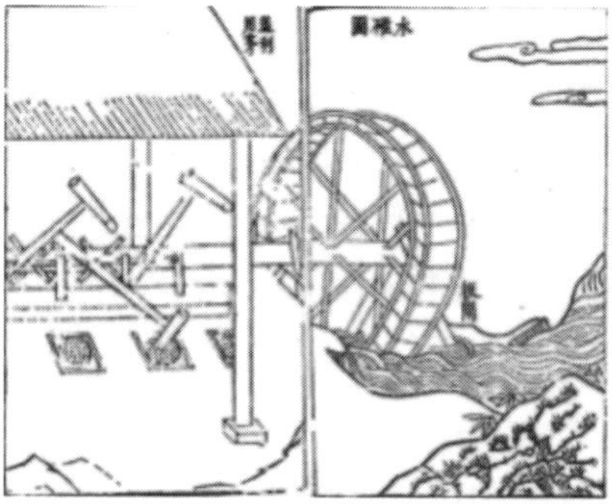

Fig.2.10 Water wheel in Min era of China

throughout the country in the 1700s in the middle of the Edo period for pumping or power.

Fig.2.11 is the oldest water wheel in Japan, which is a Uji pumping waterwheel depicted in the design of "the extract of Genji Story" written by Emperor Fushimi in the 1290s of the Kamakura periods.

Fig.2.11 Oldest waterwheel in the sketch of Extract of Genji Story (10)

Fig. 2.12 is a picture of water wheel drawn in "Syuimiyako (meaning: rare in Kyoto) Illustrated Book" in the 7th year of Tenmei (1787). It is appended that "A waterwheel is turned with the flow from the nearby valley of Kyoto Tama River. Rice is polished by using the millstone day and night to polish the

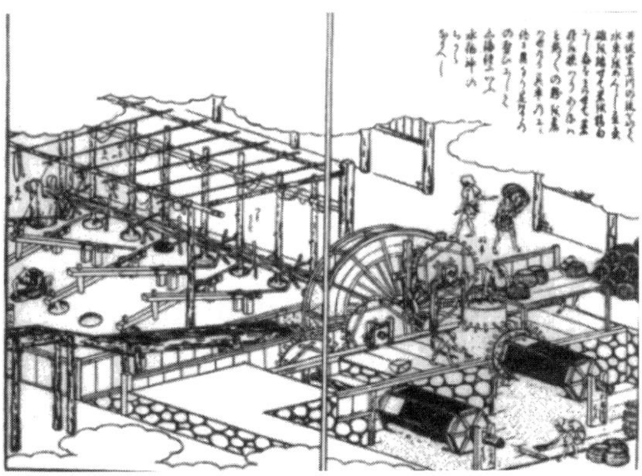

Fig.2.12 Water wheel at Tama river (Tenmei 7th year (1787) (8)

rice, to grind rape seed by rotating workman, and to create powder."

A large water wheel worked a pestle, and a gear converted the movement to extract rapeseed oil and to rotate sieves. Such moving device was rare device in Japan in those days.

The following Fig.2.13 shows the Yodo water wheel in the capital.

Fig. 2.13 Yodo water wheel (Anei 7th,1787 Issued)[8]

Hideyoshi unified Japan in 1590 because it says, "Yodo's water wheel has existed around for a long years ago, for cultivation. It was around the age." The figure depicts the Yodo waterwheel, which is famous as a pumping waterwheel, on the upper right. The figure in the foreground was a so-called 30th koku hune (ship) to Osaka. (30th koku is about 4000kg). The waterwheels were recommended to be used nationwide by Daijo-kanhu (a government) in 829, but they were not realized.

However, it seems that they were installed in the vicinity of Kyoto, and it is recorded in "Uji- Kawase Waterwheel" (by Hubokusyo) and "Ouigawa Waterwheel" (by Turezuregusa) etc., so waterwheels in Japan were realized around 1350. It means that, since waterwheels became widespread in Europe in the Middle Ages, it can be said that Japan did not fall behind so much apart from the scale. When these Japanese paintings were discovered from literature, I was deeply moved by the fact that Yamato Province was also quite good at doing it. The waterwheel that came into Japan along the "Silk Road of Flow Technologies" is thought to be ahead of qanat and next to the water clock of 660, which will be described later.

2.4 From the Industrial Revolution to the French Revolution

2.4.1 Feuds Between Newton and Hooke or Leibniz

Isaac Newton (UK,1642 to 1727) is considered to be the best physicist until Einstein's appearance in the 20th century. He is said to have been an ordinary boy with little talent in his childhood, like Edison and Einstein. At the age of 21, plague-pestilence was a pandemic in Europe and he was temporarily evacuated to his hometown. At that time, it is said that he discovered the famous "Universal gravitational force" when he saw apples fall in an orchard.

Photo 2.17 Leibniz invented Calculus (2)

However, regarding this, Hooke (Robert Hooke, UK, 1635 to 1703), who is famous for "Hooke's Law", made a complaint to Newton as “I found it first and wrote a letter to you (Newton), you got a hint from this. Therefore, you have to acknowledge this fact and publish it in the book.”

Leibniz (G. Wilhelm Leibniz, Germany, 1646 to 1716) also sent a polite letter stating that he was the first to do Calculus, but Newton ignored it, and the lawsuit continued for more than 100 years, including their successors.

It is said that this led to an international conflict between Britain and Germany, as a result, academic exchange between the two countries stopped.

Regarding calculus, Newton had no physical evidence and was at a disadvantage. However, Hooke might recognize Newton’s ability, although fighting to him. Newton announced "Principia" in 1687 with the support of Hooke who was a heavyweight of Royal Society.

Principia contains the following three volumes.

Volume 1: Newton's dynamics system, induction of Kepler's three laws from universal gravitation.

Volume 2: Motion of objects and fluid dynamics in viscous media.

Volume 3: Discussion of the space system which contains mass determination of the sun and planets and irregularity seen in the movement of the moon.

The first volume, in addition to universal gravitation, describes the first law, which is the most famous in

Newtonian dynamics, "An object at rest remains at rest, or if in motion, remains in motion at a constant velocity unless acted on by a net external force.", and the second law, "The acceleration of an object as produced by a net force is directly proportional to the magnitude of the net force, in the same direction as the net force, and inversely proportional to the mass of the object."

The best example of the first law is that an artificial satellite keeps orbiting the earth at a constant speed. In the second law, the famous following formula is derived by expressing external force; F, the mass of mass point; m, and acceleration; a.

$$F = ma \text{ ------------------ } (2)$$

Volume 2 also leaves a remarkable achievement in fluid dynamics. It is the so-called "Newton's law of viscosity" that "The shear stress is directly proportional to the velocity gradient." that is the one of basic items of fluid dynamics.

Small models and study them with water or air, for example, in experiments on ships and airplanes.

Photo 2.18 Portrait of Hook

Hooke himself has made significant achievements in

"flow measurement", with such as the anemometer and mercury barometer in Fig. 2.14.

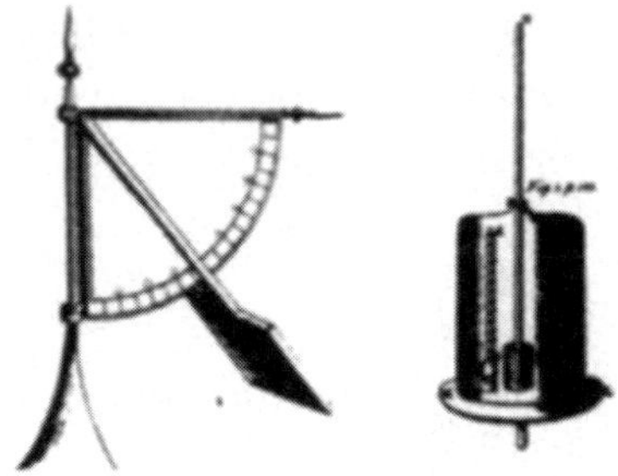

Fig. 2.14 Fook's instrument (Left: Anemometer, right: mercury barometer (17)

Dr E. A. Spencer (13) also claims that the Hooke announced a Weir flowmeter in 1683. In addition to this, he devised special instruments and apparatuses, such as universal joints and lens polishers. For this reason, it seems that he was classified as a mechanical engineer, but he was the most successful biologist in the study of biology.

Only in the latter half of the 17th century, an instrument; anemometer measuring the fluid flow appeared, which engineers involved in fluid measurement had waited for a long time.

It is also worth mentioning in fluid measurement that Newton created the concept of viscosity. However, unfortunately, Newton so disliked people because he had become tired from the battles with forerunners of inventions, and he seems to have become enthusiastic about alchemy technology in his later years.

2.4.2 Hydraulic Control During the Qing Dynasty in China (12)

Turning to China around this time, the Ming dynasty fell due to the death of King Yongming in 1662, and Emperor Kangxi took the first throne of the Qing dynasty, the last imperial court in China.

During the Kangxi, Yongcheng, and Qianlong eras, agricultural production had been very emphasized. In the last 100 years or so of those eras, not only water control of the river but also the development and utilization of water use had become somewhat effective.

A prominent contributor in this direction was Chen Huang, who had served Emperor Kangxi (1662-1722).

He invented the "water measurement method" to control floods accurately. That is, the flowrate of water was calculated by multiplying the cross-sectional area of river water by the flow velocity. According to Chen's words, "The method is to measure the size of the floodgate first, to measure how much it flows in a second next, and then accumulate the day and night to calculate the flowrate per one day."

As mentioned in Section 1.1.3, the concept of flowrate was first derived from the ancient Greek Heron around BC 150, but if the history is looked again, Chen might be the first to help the concept of flowrate that was useful to society. Why didn't such a concept come up earlier to calculate the flowrate of water, though waterwheels had been developed in Europe?

Perhaps it would be ended with a concept such as "Today, the water wheel rotates fast, so there is a lot of water." There must have been many opportunities to invent the flow meter. (Note: Turbine flowmeter appeared via Waltman in 1790.)

2.4.3 Entering the Industrial Revolution [2][3]

From the latter half of the 17th century to the first half of the 18th century, the improvement of steam engines from Newcomen to Watt, Europe rapidly progressed to the era of the Industrial Revolution.

Fig.2.15 shows steam engines made by Watt and Newcomen.

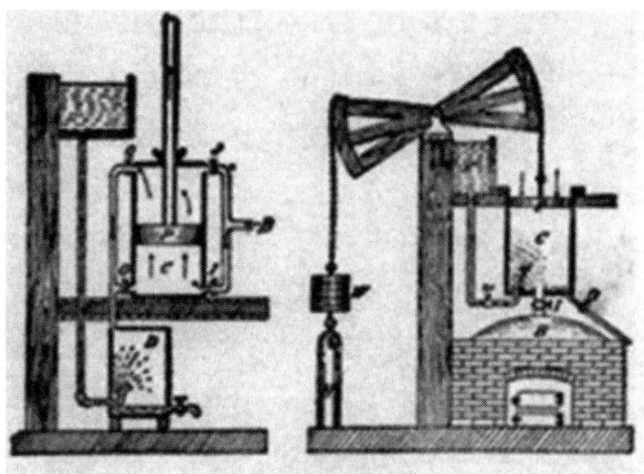

Fig.2.15 Watt's (left) and Newcomen's (Light) Steam engines [2]

Revolution centered on Britain. It was James Watt (UK, 1736-1819) who completed the steam engine and directly triggered the Industrial Revolution. Looking back at the roots of the steam engine, it is also Heron's steam reaction turbine shown in Fig. 1.20 of Section 1.1.3.

As Mr. Kesakatsu Koizumi mentioned, Heron was a very rare genius who was 1,700 years ahead of the times.

By the way, the steam engine is a decent fluid machine, but since it is endless, it will not go into the "History of flow measurement" as a heat engine or thermodynamics event.

As you know, steam engines rapidly replaced the classical power of watermill and windwheel mentioned in the previous section. Fluid dynamics was laid out by the Bernoulli school in Basel Switzerland. Daniel Bernoulli (Switzerland 1700-1782) presented in 1738 the energy principle that underlies an important fluid dynamics theorem, "Bernoulli's theorem". Based on this, in 1758 Euler established the relational expression that "the sum of the potential energy, kinetic energy, and pressure energy of the fluid is constant".

Photo 2.19 Portrait of Bernoulli [3]

Pitot (Henri de Pitot, France, 1695-1771) developed a current meter famous for the "Pitot tube" in 1732, six years before Bernoulli's theorem, and measured the flow velocity of the river Seine in this year. This is a method of detecting the flow velocity based on Bernoulli's theorem, and is still used in universities and research institutes for fluid dynamics research.

2.4.4. To the Era of the French Revolution [2][14][15][16]

When the era entered the latter half of the 18th century, following the British Industrial Revolution, the United States Declaration of Independence (1776) and the French Revolution (1789) led to a drastic change in the world. In the scientific world as well, following the discovery of nitrogen and oxygen, Lavoisier's "Chemistry Essentials", published in 1789, systematically describes the basics of modern chemistry. Since it is comparable to Newton's "Principia" and Darwin's "On the Origin of Species", it is why he is praised as the father of modern chemistry.

In this situation, a clear definition of the element concept and a scientific table of elements were presented. Galvani discovered animal electricity in 1780, and Volta discovered the battery in 1799, and finally the breath of electricity began.

The French Revolution began on July 14, 1789, when the prison Bastille, which was said to symbolize tyranny and feudalism, was occupied and destroyed by the people.

Photo 2.20 Marie Antoinette on the decapitation (14)

In 1792, the French republic was promulgated and Louis XVI was executed, after that, it entered a terrorist era where executions by the guillotine decapitation were carried out one after another.

A visit to the Palace of Versailles is highlighted by the beautiful portrait of Marie Antoinette, but she was sadly executed on October 14, 1793, by the guillotine. (Photo 2.20). Readers will be wondering why the creepy decapitation story comes up. But the reign of terror extends to scientists. The unit of length used mainly in France was Pied de Roi (about 0.325 m) until the metric system was enacted. In 1790 Talleyrand proposed to the National Assembly that a new system of units be created. The Paris Academy of Sciences also agreed with this plan and set up a committee to start a study. The members consisted of the above-mentioned Lavoisier (Antonine Laurent Lavoisier, France, 1743 to 1794), Lagrange, Laplace, and other members who had made a name for themselves in posterity. Lavoisier was in charge of measuring the density of water. Most of his work was done in August 1793, but because he was a tax collector, he was imprisoned in November of this year and then was executed by guillotine on May 8, 1794.

Photo 2.21 Lavoisier, father of modern chemistry and his wife (16)

Lavoisier went to the laboratory with an escort from prison and was executed as soon as the experiment was completed.

The Lagrange said to a friend, "It's enough to drop his neck in an instant, but such neck won't be possible even if it takes a hundred years." The density of water at 4°C announced by the successor of Lavoisier was very close to the currently most accurate water mass of 0.999972kg/dm^3

Photo 2.21 shows Mr. and Mrs. Lavoisier who were very harmonious. Marie was a blonde, blue-eyed, very beautiful lady, and helped her husband's work as "a cheerful, wise, and scientific lady", as a lifelong assistant in the salon in the laboratory. However, the guillotine brutally beheaded her.

During this period, Charles (France) introduced Charles's law on gas expansion in 1787. It is now called "Boyle-Charles's Law". He invented the hydrogen balloon in 1783. On August 24, he flew the hydrogen balloon under the watchful eye of the citizens, but failed because of an aerial explosion. Therefore, the cause of the failure was investigated. Since the balloon was improved so that it could escape when the balloon went up and expanded too much, the balloon was flown again at Versailles Square on December 1st with great success. (See Photo 2.22)

Photo 2.22 Hydrogen balloon flew at Versailles Square (16)

Meanwhile, the horror era of the French guillotine beheadings was ended with the stab of Robespierre, and the era was taken over by the hero Napoleon. It can be said that the invention of the battery by the Volta mentioned earlier was the first to light the world of electricity. It marks the beginning of electrical and electronic

technology that leads to modern scientific engineering.
After Napoleon triumphed from Italian expedition in 1796, after viewing Volta's battery experiment, he gave Volta the honor and pension of Count.
Photo 2.23 shows Napoleon's planned battle with Britain in the early 19th century, a British landing operation to cross the strait by flying a balloon developed in France from the sky, a ship from the sea, and a tunnel underground, but it was not realized.
He was destined to be completely defeated by the British Navy led by Nelson at the Battle of Trafalgar. Mitsutomo Yuasa describes the Napoleonic era as follows. "The Napoleonic era from the latter half of the 18th century to the beginning of the 19th century was the most vivid and romantic one in the history of science. The development of the mechanical worldview, which culminated in the completion of Newtonian physics, gave birth to the Age of Enlightenment in the 18th century, but it had become clear that there was an unknown vast world far beyond the field of known intellect and tradition.

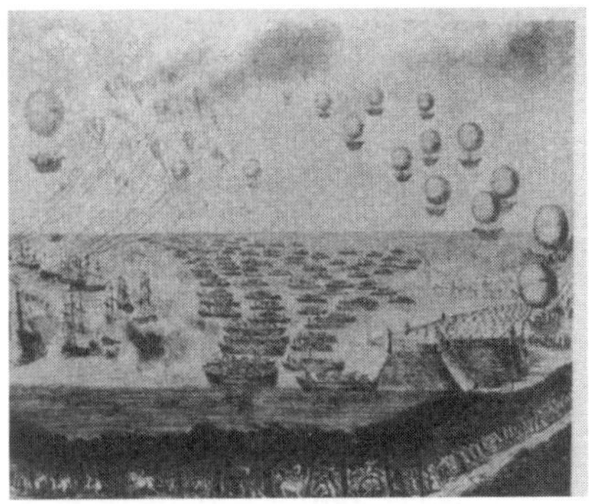

Photo 2.23 Napoleon's landing operation in England (14)

That is, it means change and advance from the mechanical view of nature, and new phenomena that could not be obtained with Newton's principle were discovered one after another. At the end of the 18th century, social institutions shifted to the era of high capitalism, and natural science achieved record-breaking development."

Reference

(1) Carlo Pedretti "Drawing of Leonardo Da Vinci/Study of nature" Iwanami Shoten

(2) Tomoasa Yuasa "Chronology of Science and Culture" Cyuou Kouronsya

(3) Yutaka Hosokawa "Cultural: Flow Science (1)" Tokyo Denki University, Press Div.

(4) Louvre Culture Department "Louvre tour guide"

(5) Popular Science volume 1,2, and 3, Grolier

(6) T.S. Reynolds "History of windmill" Heibon Sya

(7) Izumi Usiyama "Introduction to windmills"

(8) "Japanese Technology1, Draw industrial technology" Japan Society for the History of Industrial Technology.

(9) An encyclopedia of the history of technology by Ian McNeil, Routledge

(10) "Kokusi Daijiten" 8 volume Yoshikawa Koubunkan

(11) Isaac Asimov "Asimov's Scientist Biography" Syougakukan

(12) Qiu Guangming "China Ancient Weights and Measures" translated into Japanese by Jyun-itiro Kasima. The Society of Historical Metrology, Japan

(13) E. A. Spencer "Flow Standard: Past and Present" National Institute of Technology, UK, translated into Japanese by Shigeji Ishikawa.

(14) Kenji Kouno "French Revolution" Kawaide Syobou Sinsya

(15) Kesakatu Koizumi "History of Weights and Measures" Central Metrology Institute

(16) Macmillan "World Science History Encyclopedia" Hara Syobou

(17) Kan Hirata "Illustrated history of science and technology (1)

(18) From Internet Yahoo Japan

3. 19th Century Science and Technology

3.1 19th Century of the Hundred Flowers

3.1.1. 19th Century Science and Technology Civilization (1)

The era is now the 19th century. This era is the "Century of steam" from the viewpoint of industrial technology, but it is an era in which the scientific and technological foundations for moving to the "century of electricity" in the 20th century were rapidly being completed. The enactment of the metric system and the founding of École Polytechnique during the French Revolution were great legacies for the next century. The metric system was decided upon in 1790, stating that 1 meter is 1 of 40 million points of the meridian length passing through Paris. The École Polytechnique was founded in 1794 as a school of artillery or crafts. Napoleon said that it was a training center for "Hens that lay golden eggs." In the first quarter of the 19th century, famous mathematicians and physicists such as Laplace, Lagrange, Carnot, and Fourier served as instructors, and their education produced excellent scientists and engineers.

In the next quarter century, Engels for natural dialectic, Faraday discovered the fundamental principle of electrical energy generation, and Morse, famous for the "Morse code" began to play an active part. In the latter half of the quarter century, Darwin published "On the Origin of Species" and Mendel did "The Law of Inheritance", that had contributed to the development of biology. In addition, Liebig who was the father of organic chemistry appeared and laid the foundation for current organic compounds. The theory of the most important concept "field" since Newton appeared, and Maxwell completed "electromagnetism". In the last quarter century, Edison of the United States invented a carbon beam bulb. In addition, medicine began to move rapidly, Koch discovered tubercle bacilli and cholera. Typhoid fever and diphtheria were discovered one after another, and Kiyoshi Shiga of Japan also discovered Shigella in 1897.

French science with the horde of talented scientists from École Polytechnique in the first half of the 19th century and boasted global advantage lagged behind in its technical aspects, in the 1880s, as it was overtaken by the underdeveloped countries of America and Germany.

Comparing the status of major countries in world industry, Order became,

1860s --------- UK, France, US, Germany

1900s --------- US, Germany, UK, France

From a technical point of view, the internal combustion engine had advanced, and Daimler (Germany) built a gasoline engine for automobiles, and succeeded in driving the first car in 1883.

In terms of chemical industry technology, in the early 19th century, coal gas lamps were lit in the order of London, Paris, Vienna, and Berlin, and in 1859, oil fields were excavated in Pennsylvania in United States, and also the oil industry broke out. The production and sale of gas and oil would require instruments that display the amount in real time, that is, flow meters.

3.1.2 Inventions and Discoveries of the 19th Century

Then, let's list the inventions and discoveries of the 19th century that had great influences on the flow measurement of posterity.

(1) Experimenter Faraday [1][2][3][4][5][6][7]

Electrical energy gave huge impetus to scientific civilizations in the 19th century along with thermal energy.

Faraday (Michael Faraday, 1791-1867) established the fundamental principle to use electrical energy.

He was born in 1791 to the son of a poor blacksmith. He was sent to an apprentice at a bookbinder at a young age, and became very interested in science owing to contact with books there.

Photo 3.1 Faraday's portrait [5]

The owner of the bookbinders encouraged him to read and allowed him to attend science lectures. He mostly enjoyed the lecture by Humphry Davy, who was one of Britain's most renowned chemists at that time. After taking notes carefully, he returned to home and carefully made a clean copy in addition to charts and memos. Faraday sent the notes to Davy and offered to be an assistant in his lab. Davy was astonished by his notebook and hired Faraday as his assistant as soon as his assistant post was available. At the age of 22, Photo 3.2 is a picture of Faraday experimenting with Davy.

Davy has made various inventions and discoveries, but the greatest discovery is said to be "To find Michael Faraday." Faraday discovered the phenomenon of electromagnetic induction in 1831. Based on this, he invented a generator composed of magnets and coils,

Photo 3.2 picture of Faraday experimenting with Davie (7)

Photo 3.3 Faraday's invented generator (5)

Photo 3.4 Faraday's experiment on the Thames (6)

and succeeded in turning it to continuously generate electricity (see Photo 3.3). Modern society is receiving the most benefits.

In addition, the principle of electromagnetic induction is also the source of Electromagnetic flowmeter that are in great use today. There are a following sections about it. He attempted to measure the flow of the Thames using electromagnetic induction. He hung a 960-foot (296.2m) long copper wire from Waterloo Bridge and dropped another copper wire at the tip with a metal plate that was submerged in the river a little bit.

A report submitted to the Royal Society in 1832 found that the ammeters fluctuated. As a matter of fact, the riverbed broke the signal of flow, and the signal was greatly shaken due to the disturbance, so that his initial purpose could not be achieved after all. However, he fortunately lived long enough that he heard that in 1851 Walston succeeded in measuring the tide level of canal with an induced potential measurement.

By the way, it was Faraday was the first to liquify gas.

He sealed a boomerang-like bent tempered glass with a substance that easily produces gas when heated. When the glass tube was immersed in boiling water, the tube filled with gas and pressure increased. When the opposite end was placed in a beaker filled with ice, the gas was placed in a low temperature state at high pressure, as a result liquefaction happened. In this way, chlorine was successfully liquefied in 1823. He also liquefied several other gases. He conducted a wide range of experiments not only in electromagnetism, but also in fluid dynamics and thermodynamics, that achieved good results. He was an indeed characteristic experimenter. He couldn't use mathematics like Newton for the rest of his electromagnetic fields.

In addition, the principle of electromagnetic induction is also

Einstein established the “Special Theory of Relativity” in the 20th century, but he said that he was still no good at mathematics so, did not use advanced mathematics so much. Therefore? It is said that a portrait of Faraday was hung on the wall of Einstein's library.

(2) Coriolis Force (4)

In 1835, French engineer Gaspard Gustave de Coriolis (France, 1792–1843) solved Newton's second law in a rotating coordinate system and found that a centrifugal force and an apparent power called the “Coriolis force” works in moving mass points.

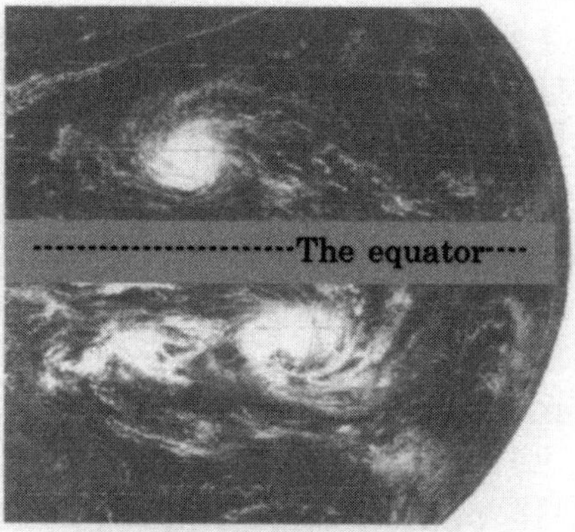

Photo 3.5 Two storm directions on Earth by Coriolis force

This is the principle of the Coriolis flowmeter, which is currently called the latest "true mass flowmeter". As a prominent example, as shown in Photo 3.5, the Earth rotates from left to right, so a storm-vortex occurring on the Earth turns clockwise in the Northern Hemisphere, but it turns counterclockwise in the

Southern Hemisphere at the equator. This phenomenon is caused by the Coriolis force.

(3) Doppler Effect (3)(4)

With the advent of steam locomotives, humans began to

experience interesting phenomena that they had never noticed before. It was the phenomenon that the tone of horn sounds higher when the locomotive approached and lower when the locomotive moved away.

Photo 3.6 Doppler's portrait [(4)]

Austrian physicist Doppler (Christian) Johann Doppler, 1803 ~ 1853) said, "Where the sound wave moves with the sound source, when the sound source approaches a human, the interval between the sound waves that reach the ear becomes shorter (the frequency becomes higher), that is, the timbre becomes higher. On the contrary, when the sound source moves away, the interval between sound waves becomes longer (the frequency becomes lower), therefore the timbre becomes lower."

In 1842, Doppler had confidence in this interpretation, and two years later he experimented with his theory. For two days, a freight car without a roof or a wall was pulled by a locomotive while changing the speed, and the sound of the trumpet played on the freight car is heard by a music troupe with excellent sound on the ground, and the tone changes depending on the speed recorded. Perhaps it was recorded that the sound of C changed to C # or C*b* depending on the speed and direction of the locomotive.

In this way Doppler proved the correctness of his theory. The Doppler effect played an important role in astronomy a few years later. This is also the principle of the current Doppler type ultrasonic flowmeter.

(4) Absolute Zero [3]

Since Amontons conducted an experiment to reduce the volume of gas with temperature in 1699, there was a theory that the temperature of liquid would decrease infinitely, but in 1848, William Thomson (UK, 1824 ~1907) who later became Baron Kelvin assumed that -273 ° C (currently -273.15 ° C) was absolute zero, and proposed to set a new temperature scale with absolute zero as the zero point.
Therefore, this absolute temperature unit has come to be represented by K which is the initial for Kelvin. Since the volume of gas expands and contracts according to the ratio of absolute temperature, it is important for measuring the volume of gas.

(5) Mach Number [4]

Austrian physicist Ernst Mach (1838-1916) published a study in 1866 on the phenomena caused by hard objects and air moving at high speeds against each other. The speed that air molecules can move is the speed that sound travels through the air. When an object moves at high speed in the air, the air molecules can no longer naturally follow the object, they are pushed aside by the object, and the air molecules are faster than they want to move, and a new situation creates. For example, supersonic motion compresses air and makes an impact sound when it expands. The best example of this is thunder, that is the

Photo3.7 Portrait of Mach [4]

impact of the heat of lightning causing the air to expand faster than the speed of sound.

It is said that the whip of skin also makes a beeping sound because the air is faster than the speed of sound. Currently, in honor of Mach, the speed equal to the speed of sound is called Mach 1, and the speed twice as fast is called Mach 2.

(When I was in junior high school, I watched an American Movie about Mach. It was a scenario where aviators tried to break through Mach 1 by diving jet planes from the sky. A hero of aviators broke through Mach 1 for the first time, after overcoming the tragedy of comrade who brutally crashed into the ground. I remember seeing it while holding my breath. The voice of the aviator yelling "Vibration! Vibration!" when crossing Mach 1 still remains in my ears.)

(6) Reynolds Number [4]

A great British Hydrodynamicist; Osborne Reynolds (1842-1912) demonstrated his outstanding talent as an experimenter, which gave him the most basic concept in fluid dynamics. He taught at a college (later the University of Manchester), initially conducting research in a wide range of fields from physics and engineering, such as electromagnetics and thermodynamics, but began researching fluid engineering around 1874. After that, he conducted the famous

Photo 3.8 Reynolds's portrait [4]

experiment of observing laminar flow and turbulent flow in a pipe, and established the Similarity Law.
If V is the velocity of object moving in the fluid, D is the representative length of object, and ν is kinematic viscosity coefficient of fluid, the following equation holds.

$$\mathrm{Re} = \frac{V \cdot D}{\nu} \quad \text{-----------------------------} \quad (3)$$

A dimensionless number shown by can represent flow state. Re is the Reynolds number, which is a value that determines whether it is laminar flow or turbulent flow, and is an indispensable parameter for fluid analysis. Even in the current flowrate measurement, it is an indispensable parameter especially for the inferential type flow meter (differential pressure type, turbine type, Karman vortex type, etc.). In this way, in the 19th century, the foundation for real flow measurement was almost finally completed.

3.2 History of Science and Technology in Japan

3.2.1 History of Science and Technology Up to the 19th Century in Japan (1)

So far, the history of science and technology in the West has been the focus, but let's take a look at Japan.
The epoch in the history of science and technology in Japan would be,
1543 The introduction of guns
1639 The ordinance of Closed Country (Sakoku)
1720 Foreign books lifted
1868 Meiji Restoration

-Before the introduction of guns, civilization and culture were mainly introduced from Sui, Tang and Kudara, and while Buddhism became widespread, they learned architecture and crafts in particular, which developed into the construction of shrines, temples and Buddhist statues unique to Japan.

-Under these circumstances, what should be noted as the history of science and technology is that a solar eclipse was recorded in 628, a comet was recorded in 634, and Nakano-Oue Prince made a Roukoku (water clock) in 660 in order to set the time system (time Origin), coal (burning soil) and oil (burning water) were discovered in Echigo in 668.

-After the Portuguese arrived at Tanegashima and introduced guns, Spaniard Francisco de Xavier and British Adams (Japanese named; Miura Anjin) came to Japan. So, Japan came into contact with Western science and technology.

-It is well known that guns and cannons became widespread during 100 years from the introduction of guns to the Closed Country, and person who successfully adopted them during the Warring States Period began to make themselves master of the Empire.

-Western typography was introduced in 1590, 140 years after Gutenberg invented typography in 1450.

-In 1605 Adams built two Western-style sailing ships in Ito. (The beginning of Western-style shipbuilding).

-Nobunaga Oda (~1582) was quite enterprising and tolerant of Christian missions, but after starting with the ban on Christians by Hideyoshi Toyotomi (~1598), trade with countries other than the Netherlands and China was banned by the ordinance of Closed country of Tokugawa Shogunate.

-As a result, until Yoshimune lifted the ban on foreign books in 1720, a small number of Western studies would come in from Dejima in Nagasaki through Dutch.

-In the meantime, Seki Takakazu (1642-1708) made great achievements in Wasan (Japanese arithmetic). In addition, the Tokugawa Shogunate encouraged agriculture, so agricultural technology was advancing.

-After the ban on foreign books was stopped, Ryotaku Maeno and Genpaku Sugita published the famous "Kaitai Shinsho"(for medical science) in 1771, and Gennai Hiraga made "Elekitel" in 1776. Tadataka Ino started surveying all over Japan in 1800 and made the first map of the country.

-Philipp Franz von Siebold came to Japan as an employee of the East India Company in the Netherlands in 1823 and had a great influence on the history of science in Japan. It must be seen that the real transplantation of genuine science such as physics, chemistry, and mathematics, which are the foundations of modern technology, began after Siebold's arrival in Japan.

-Japan's first physics book was "Kikai Kanran" written by Rinsou Aochi in 1825, and first chemistry one was "Seimi Kaisou" written by Yoan Udagawa in 1837.

-In 1858, Sakuma Shozan made permanent magnets and Daniell cells.

-Since the arrival of US Ambassador Perry in Uraga in 1853 with four black ships, the transition occurred with the collapse of the Edo Shogunate. The period dramas continued through the restoration of the royal government, to the civilization of the Meiji era, one after another, and so Japan rapidly westernized.

3.2.2 Roukoku (Water Clock) [8][9][10]

Photo 3.9 Portrait of Nakano-oue Prince [14]

In “Nihon Shoki”, it was written that in May of 6th year of Emperor Saimei (660), “The crown Prince made a Roukoku for the first

In “Nihon Shoki”, it was written that in May of the 6th year of Emperor Saimei (660), “The Crown Prince made a Roukoku for the first time and let the people know the time."

The crown prince was Prince Nakano-oue (626 ~ 671), who later became Emperor Tenji. The Roukoku was a device that measured time based on the amount of water in an aquarium that dripped through a small hole.

Photo 3.10 The ruins of a water clock made by Nakano-oue Prince [9]

It is believed that the structure (Photo 3.10) found at the Mizuochi Ruins in Asuka Village, Nara Prefecture is probably this ruin. In the photo on the right, the Japanese-lacquered wooden box on the lower side and the water supply/ drainage ditch on the upper side can be seen. In addition, there is a trumpet-shaped copper pipe in the upper part of the left photo, and a wooden gutter for water conveyance in the lower part Fig. 3.1 may be a picture of a water clock made by Prince Nakano-oue according to the "water clock theory" written by Yosen Sakurai in 1748. Water flowed from an upper wooden box to a lower wooden box at a constant flowrate, and the time might be known from the amount of water. It can be said that this is the first flow control in Japan to measure the time accurately.

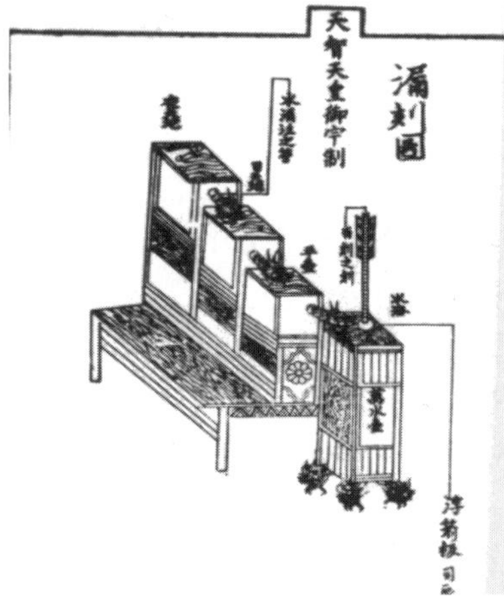

Fig.3.1 Water clock of Nakano-oue prince by Yozen Sakai (10)

Figure3.2 is an imaginary view of the time signal system at Yamato court that Roukoku was set on the first floor, and a bell and a drum were set on the second flower to inform the time to people.

Fig.3.2 Time signal system of Yamato (8)

In Tang, there was a similar work by Lu Cai, who was active around 627-649, and it is considered that this was introduced to Japan

3.2.3 Electrical Technology and Measurement in the Edo Period [10][11]

Photo 3.11 Portrait of Gennai Hiraga [11]

Gennai Hiraga is said to be a pharmaceutical scholar and a gisaku (popular)/Joururi (storyteller) writer, but he is known as the first science and technology engineer to create an electric Elekiteru in Japan. Elekiteru is a transliteration of the Latin word "Electriciteit". Gennai Hiraga in photo 3.11 got a broken Elekiteru in Nagasaki. It is said

Photo 3.12 Elekiteru made by G. Hiraga [11]

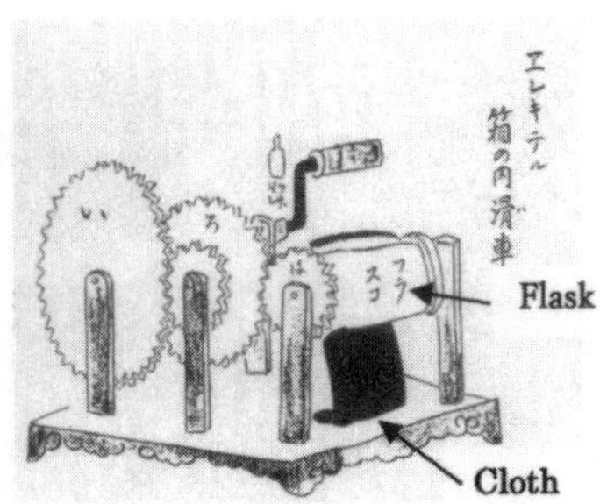

Fig.3.3 Principle diagram of Elekiteru [10]

that he took it back to Edo in 1776 and spent seven years restoring it. Elekiteru is a friction generator in the 1740s. The static electricity experiment was popularized in the West by Frenchman Jean Antoir Nore and others, and it came to Japan through the Netherlands. Photo. 3.12 shows the Elekiteru, that is said to be made by Hiraga, which is

currently in the Ministry of Communications Museum. Then, what was happening to this structure was detailed in 1811 by
Soukichi Hashimoto's “Research of Holland-made Elekiteru”. Fig. 3.3 shows the principle of an Elekiteru. When the handle is turned, the glass flask and cloth rub against each other via a gear transmission device to generate static electricity.
When the electric wire at the top of Photo 3.12 is touched with both hands, electric current runs.
In addition, Fig. 3.4 is a "Figure of a hundred people" drawn in the above book by Soukichi Hashimoto.

Hundred people connecting by their hands **Elekiteru**

Fig.3.4 Figure of surprised hundred people (10)

Explaining this figure, there is an electric wire in a room behind sliding doors, and an electric wire is linked to two knobs rear of the fusumas (Japanese sliding doors).

In the front room, two persons touch the knobs. Many 100 people are handling in a circle with them. When the person behind of fusuma turns the steering wheel of the Elekiteru to generate static electricity, the electricity runs and the people are surprised to feel it, and if surroundings are dark, sparks are emitted. It is stated that doctors and others also paid attention to the intense electric shock and tried to treat paralysis and rheumatism.

The technology of static electricity was introduced from the West to Japan with a delay of about 30 years, but after Hashimoto, there was no progress in electricity like Faraday in Europe.

In 1768, Gennai Hiraga produced a thermometer as shown in Fig.3.5 in cooperation with Junan Nakagawa. In this way, Hiraga was the prominent engineer at that time, but in his later year he accidentally killed two people and was imprisoned in a prison in Kodenma-cho. One month later, on December 18, 1779, he died there having contracted tetanus.

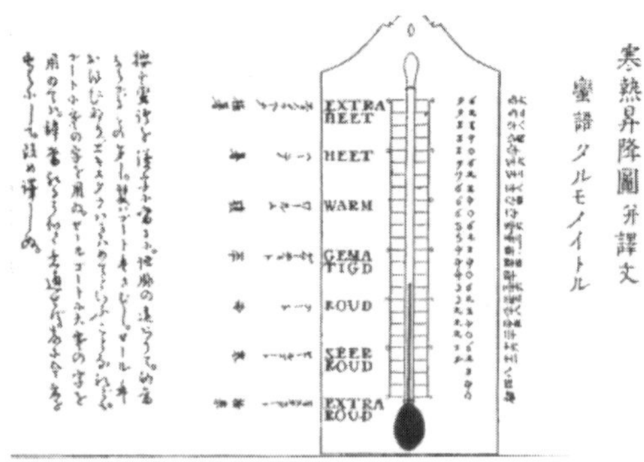

Fig. 3.5 Thermometer by Hiraga (10)

Though he used to serve the Edo Shogunate, he discovered minerals and medicinal stones, and had a friendship with Genpaku Sugita, who mourned his death. He invested his own money to build a tomb table and left the following poem for Hiraga.

"*You weren't an ordinary person!*
Like non ordinary things and actions weren't ordinary.
So, this was really an emergency death! "

Yoan Udagawa put a buoy type hydrometer like Photo. 3.13 in "Shamitu Kaisou" in 1837. In the figure, 1 is for universal use, 2 is for measuring a liquid lighter than water, and 3 is for measuring a liquid heavier than water. Next to it a note was written. Universal use (1) came from Amsterdam, but it was inconvenient because so long, therefore 2 and 3 were made.

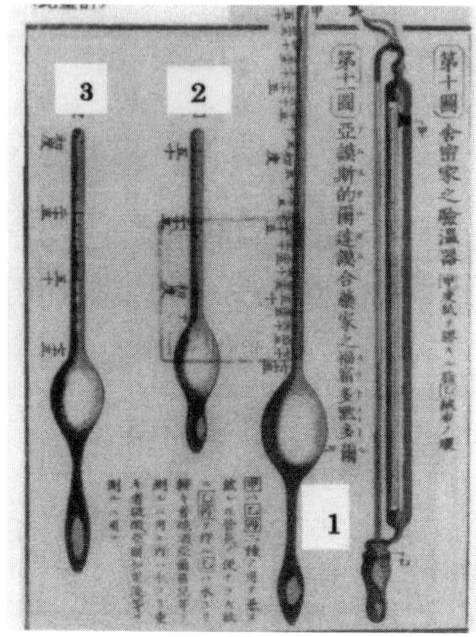

Photo.3.13 Hydrometer in Edo Era (10)

3.2.4 Pumping Machine in the Edo Period (10)(12)

In 1836, Nariaki Tokugawa compiled "Ungeikisan", and classified about five pumping machines, as "Kikkousho", "Shoukouryu", "Kensensya", "Kyusuitou" and "Honryusya" as shown in Fig. 3.6.

The internal structure of everything cannot be understood, but maybe:

(a) Kikkousho is designed to be a foot suction pump.
(b) Syoukouryu is designed to use a few buckets inside.
(d) Kenseisya and (e) Honryuusya are using impellers.
(c) Kyuusuitou looks like a Ryuo wheel (Archimedes Screw) as shown in Fig. 3.7.

What an ancient Greek technique was transmitted to Japan!

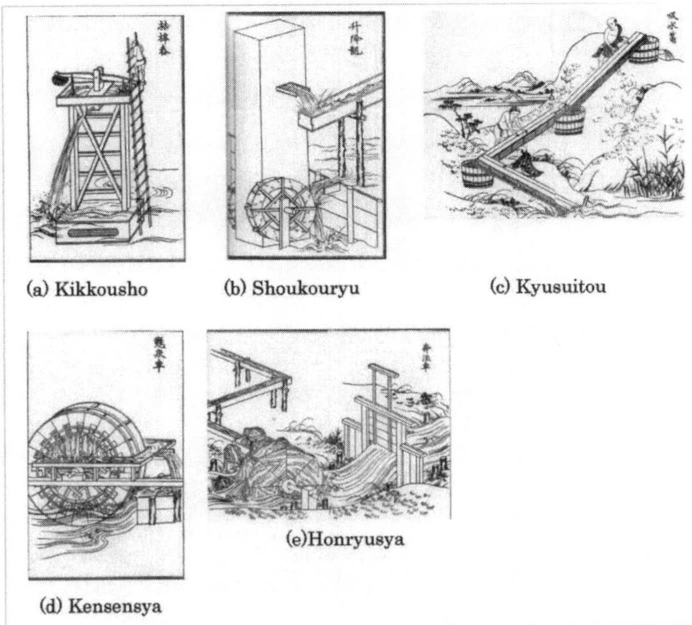

Fig.3.6 Classification of pumping machine in the Edo Period [10)]

Fig.3.7 Archimedean spiral Water Pump (10)

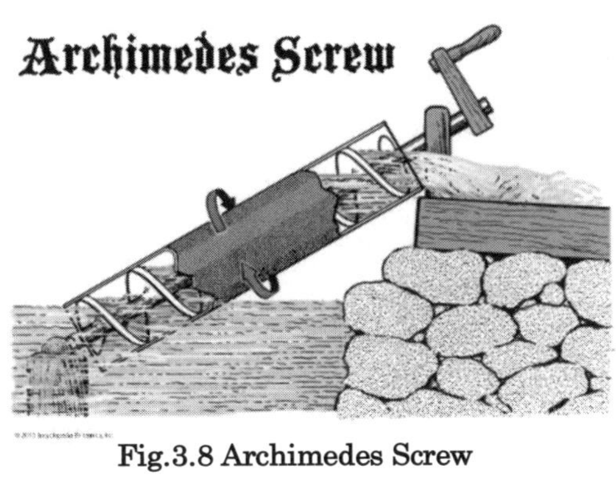

Fig.3.8 Archimedes Screw

Pumping Technology) and also (See Fig.1.28 in Section1.2.3 recently, concerning Archimedes Screw, a photo (Fig.3.8) was found on the internet. The Archimedean screw, or Egyptian screw, is a machine used for transferring water from a low-lying body of water into irrigation ditches. Water is pumped by turning a screw-shaped surface inside a pipe. The Archimedes screw shown in Fig.3.7 traveled all the way from Egypt to Japan along the Silk Road of flow technologies!!

Fig. 3.9 Many Archimedean spiral pumps used in "Sado Kinzan Kanahori no Zu" in Edo era (Cabinet Library Collection) [10]

The center figure of Fig.3.7 is part of the screw, and the upper one shows the whole Archimedes Screw and the one below is complete with a base
The next application (Fig 3.9) is a real one in Japan.
Fig. 3.9 shows "Sado Kinzan Kanabori no Zu". This means “The figure of digging gold at Sado Golden Mountain.” (In Edo Era this was owned by the Japanese Cabinet Library). In total 5 Archimedes Screws were used to pump up water.
Fig.3.10 is an explanatory diagram of the steam engine by Newcomen and Watt, which was described in "Ensei Kikijutu" written by Koumin Kawamoto in 1854. It was introduced to Japan 150 years after the improvement of Newcomen's steam engine.

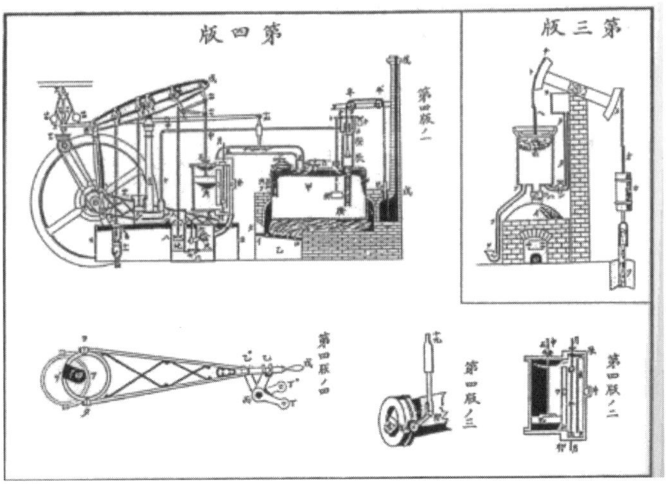

Fig.3.10 Explanatory drawing of steam engine in Ensei Kikijutu (10)

Kesakatsu Koizumi (March 10, 1918-1998) was a Japanese metric history researcher who contributed to weights and measures and measurement research. In this book, his writings were quoted elsewhere. Regarding the measurement technology of the Edo period, he said, "As for the technology of weighing and metrology in Japan of this era, there is only a little remarkable development in a special field such as surveying technology and surveying

Fig.3.11 Customs related to weights and measures in the Edo period (10)

instruments used by Tadataka Ino. When looking at surveying technology as a whole, it progressed little from the framework of medieval technology."
(Note: Tadataka Ino; It took 17 years, from 1800 to 1816, to survey all of Japan and compile the first complete Japanese map to clarify the exact appearance of the country.)

Under such circumstances, Fig.3.11 shows the customs related to the weights and measures in the Edo period.
(a) a money exchanger, (b) a weigher, (c) a copper storefront, and (d) a scale-maker.

3.2.5 Overture to Modern Science (12)(13)

The politics of the Shogunate were collapsed by Taisei Hokan (Returned the government to the emperor) in 1867, and the era moved to the Meiji Period of civilization.

The telegraph started between Tokyo and Yokohama in 1869. In 1871, with Tomomi Iwakura as the plenipotentiary ambassador, the Iwakura Mission that included the brilliant members Toshimichi Okubo, Takayoshi Kido, Hirobumi Ito and others visited the United States and Europe for one year and ten months. Kunitake Kume (Photo 3.13), a follower among them wrote "Real memories in travel around United States and Europe" from the scientific angle. It described in detail the weights and measures and industrial technology, including

Photo 3.13 K. Kume (13)

physics, chemistry, hydraulics, and thermodynamics in the United States and Europe.

From then on, there were some false explanations concerning dynamics and hydraulics, but his achievement was great and surprising to see. Photo 3.15 is an elemental table introduced in the Kume collection "Kagaku Kangen (Chemical book)." The atomic weight of Youki (oxygen) is 8, and Keiki (hydrogen) is 1, etc., which were written in Chinese numerals.

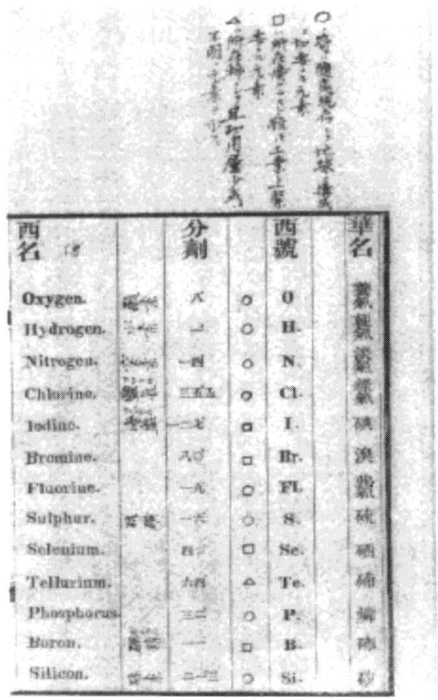

西名		分劑		西號	華名
Oxygen.	[illegible]	[illegible]	○	O	[illegible]
Hydrogen.	[illegible]	[illegible]	○	H.	[illegible]
Nitrogen.	[illegible]	[illegible]	○	N.	[illegible]
Chlorine.	[illegible]	[illegible]	○	Cl.	[illegible]
Iodine.	[illegible]	[illegible]	□	I.	[illegible]
Bromine.		[illegible]	□	Br.	[illegible]
Fluorine.		[illegible]	○	Fl.	[illegible]
Sulphur.	[illegible]	[illegible]	○	S.	[illegible]
Selenium.		[illegible]	□	Se.	[illegible]
Tellurium.		[illegible]	△	Te.	[illegible]
Phosphorus.		[illegible]	○	P.	[illegible]
Boron.	[illegible]	[illegible]	□	B.	[illegible]
Silicon.	[illegible]	[illegible]	○	Si.	[illegible]

Photo.3.15 Elemental Table (13)

In 1897, the first railway opened between Shinbashi and Yokohama was known for the song "Kiteki issei Shimbashi wo---- (From Shimbashi by one steam whistle---)", so, rapid modernization began in Japan. (See Photo3.16)

Under these circumstances, the Mathematical and Physical Society of Japan, Chemical Society, Engineering Society, etc. were established around the 10th year of the Meiji era, and a proprietary patent ordinance was proclaimed in the 18th year of the Meiji era (1885).

From the standpoint of measure, Japan joined the Meter Convention in 1885. As describing in Section 2.4 "From the

Photo 3.16 Japan first locomotive From Shimbashi to Yokohama

Industrial Revolution to the French Revolution", it had been about 95 years, since Talelan proposed the establishment of the General Conference on Weights and Measures in 1790, during the French Revolution. The treaty was promulgated on April 16, 1886 by Prime Minister Hirobumi Ito and Foreign Minister Kaoru Inoue.

Photo 3.17 is the first standard shaku and kan made in Japan by order from the International Bureau, and was received in October 1889. Then, in 1893, the shhakkan-ho was enacted using a metric prototype.

(Note; 1 Shaku = 0.303m, 1 Kan = 3.75 kg)

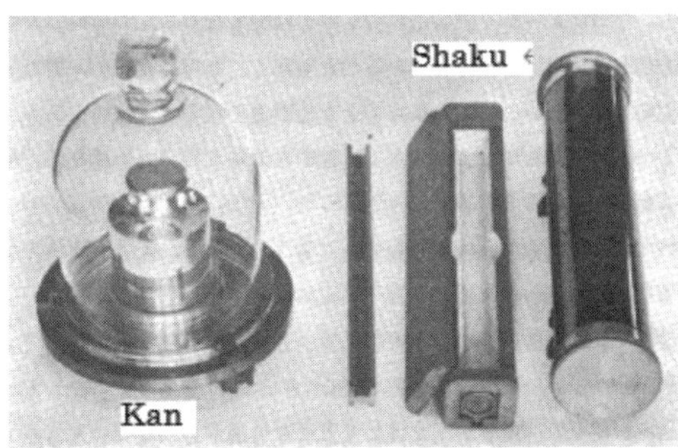

Photo 3.17 The first standard device for shaku and kan (12)

The history of science and technology in the world and Japan has been talked in a redundant manner, but it was thought that "the

history of flow measurement" could not be talked without the history of peripheral science and technology.
However, it seems that flow measurement is ripe for talking right about now, so now it is finally time to get into the main subject of the next section

Reference

(1) Mitsutomo Yuasa "Chronology of Science and Culture History" CYUOKORON Inc.

(2) Isaac Asimov "Asimov's Scientist Biography" Shougakukan Inc.

(3) Isaac Asimov "Chronology of Science and Discovery" Maruzen Inc.

(4) Yutaka Hosoi "Cultural and Flow Mechanics (1)" Tokyo Denki University Press

(5) Kan Hirata "History of Illustrated Science and Technology"

(6) E. A. Spenser, Translated by Shigeji Ishikawa "Flow Standards Past and Present" UK National Institute of Technology

(7) "Popular Science" volume 5, Grolier

(8) Hideyuki Kasahara "78 people lived in history" PHP Institute, Inc.

(9) "Japanese History Encyclopedia" Yoshikawa Koubunkan

(10) Toshihiko Kikuchi "Picture Book: Edo Period Technology (1) (2)" Tsunewa Publication

(11) Isamu Shirofuku "Hiraga Gennai" Yoshikawa Kobunkan

(12) Kesakatsu Koizumi "History of Weights and Measures" National Institute of Advanced Industrial Science and Technology Central Metrology Inspection Center

(13) Seiji Takada "Scientific Spirit of the Restoration" Industrial Technology seen in "American and European Circulars" -Asahi Sensyo

(14) Japanese History "Person encyclopedia" Shueisha Inc.

4. Development of the Flowmeter

4. 1. History of Flowmeter Development

4.1.1 Roots of Flowmeter Development (1)(2)

Exploring the roots of flowmeter development, as described in "1. Relationship Between Water and Humankind, 1. 1 Ancient Orient", it is said that it started with the measurement of flowrate of the Nile River, which was measured several thousands of years ago with a water level gauge (Nilometer). Also, in ancient Rome, Heron (Greece, 130-75 BC) proposed that the flowrate is "flow velocity multiplied by cross-sectional area of the flow path", therefore the daily quantity of water by the Roman aqueduct was calculated as 990,000 m^3 first. However, how the flow velocity was measured is not clear within the range investigated. A clock is needed to calculate the flow velocity. Sundials were used in ancient times, but Ctesibius (Greece) of Alexandria invented the water clock in 245 BC. In 159 BC, the Roman Scipio Nasica imported the water clock into Rome. Therefore, in the days of Heron, it was probably possible in Rome to pour a tracer as light as a leaf into the waterway (L), and measure the time (t) required by a water clock, and calculate the flow velocity (L / t).

Formerly a NEL (British National Institute of Technology) flow measurement lead scientist, Dr. Spencer (E.A. Spencer) described the ancient flow measurement technology as follows: "The theories of philosophers like Plato and Aristotle were descriptive and in contrast to the realistically developing Roman culture, and their calculations were quite simple. The talent of Roman engineers demonstrated during the construction of the waterworks was important, even if their theoretical knowledge of fluid dynamics was not systematic. Long before Castelli (Benedetto Antonio Castelli) in Italy ideas were systematically established as a general theory in 1639, Heron in the first century BC had come up with the idea of finding the passing flow from the flow velocity and cross-sectional area, therefore, it can be said that the Greeks theorized it and the Romans systematized it. Agricultural irrigation water was developed by the authority of the Roman emperor, and water supply systems were established in all large cities. Speaking of Rome, at the end of the 1st century, a technician name Sextus Julius Frontinus (40 AC~103 AD) wrote the following overview.

For the price of water flowing continuously from the water pipe, the pipe size was used as a gauge for measuring the amount of water supplied, but since it was easy to be fraudulent, it was replaced by the method that inserted a brass tube into the individual supply pipe. It was 1 foot (about 25cm) long and had a structure divided into many 5 divided areas. *(Author: It is thought that one of the resistance elements was devised so that a constant flowrate would flow.)* It can be said that this was the first flow measurement standard. In the latter half of the 17th century, the French Mariotte invented a pentagon-like pauce as a measuring instrument for drainage. He measured the flowrate from the pauce with an opening of 1 inch when used with different heads and tabulated this. This theory

was finally established by Bernoulli in 1738, about 50 years later.
Mariotte showed that the flow velocity in the open culvert was slow as it approached the bottom. He demonstrated this by placing multiple floats (flexibly) connected to each other at certain depths.

Photo 4.1 Open-ditch flow

In addition, the Weir flowmeter was announced by Hook in 1683 for open-ditch flow, and as it was used effectively until the end of the 18th century, open-ditch flow measurement was much earlier than in-pipe flow measurement. Concerning the price of gas supplied to homes in England in the 18th century, was determined by the number of gas lamps in the house rather than the actual consumption, and the amount of water supply was determined by the size of house. The first gas meters were made in 1815. When it comes to water meters, in the UK these are still now used only in very limited areas, probably because of its abundant water resources."
Table 4.1 shows the history of flowmeter development created by Dr. Spencer. The table is divided into the initial R & D stage, product commercialization, national standard, and international standard, but it is added that this shows the UK national standard.
After this, the history of various flow meters will be described, but please refer to this table, concerning the position and future development of each type of flowmeter.

Table 4.1 History of Flowmeter Development

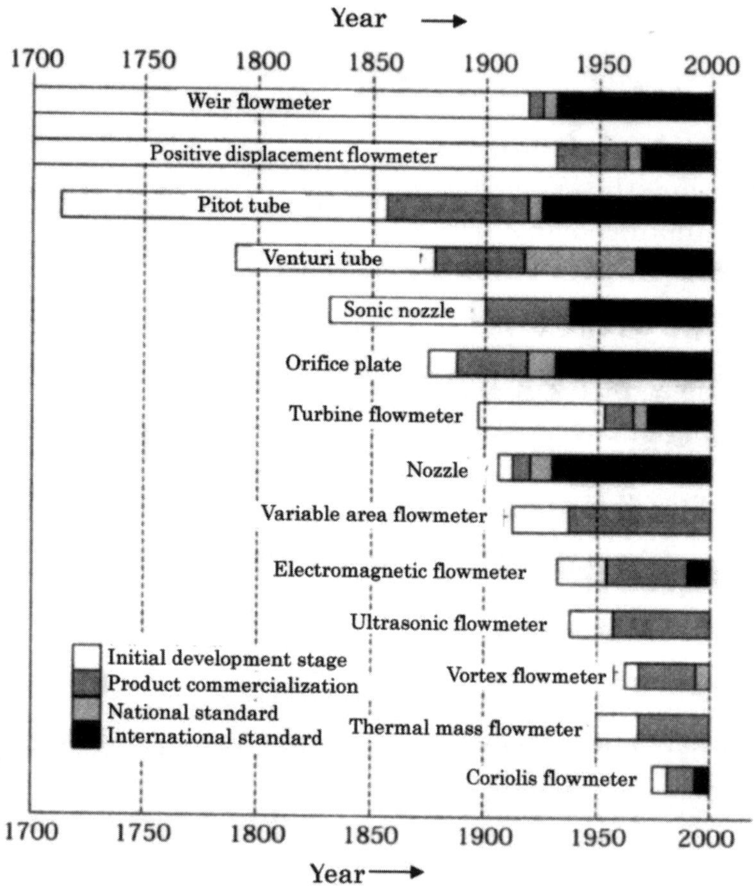

(1) Weir Flowmeter (4)(7)

According to Dr. Spencer, the first flowmeter in the early research phase was the weir type flowmeter for open-ditches (Refer to Photo 4.1). The current weir type flowmeter is shown in Fig. 4.1. When the flowrate is small, the triangle weir as shown in (a) is useu, but with a larger flowrate, the square weir (b) or full-width (c) are used.

Flowrate is calculated by the dimensions of the alphabet indicated in the figure, also the flowrate changes depending on the dimension of height (h). Therefore, the water level h is an important parameter.

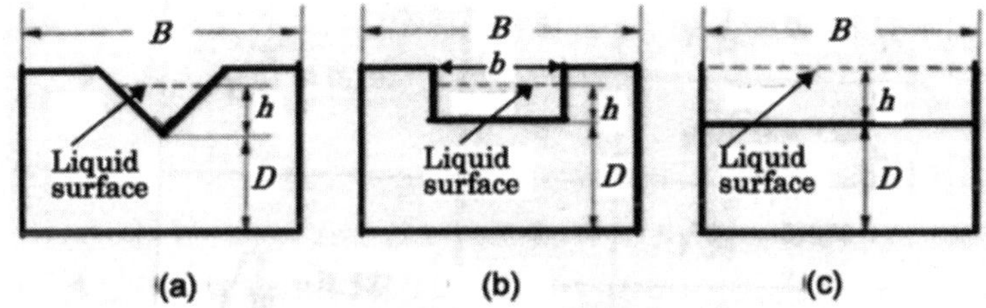

Fig. 4.1 Principle diagram of weir [7]

The hook gauge as shown in Fig.4.2 was devised in 1840 by Hoyden of Boston. This is a method in which a needle is lifted from bottom to top, and at the moment when a needle tip protrudes to liquid surface it catches the spread as a wave of light because of the momentary change in surface tension.

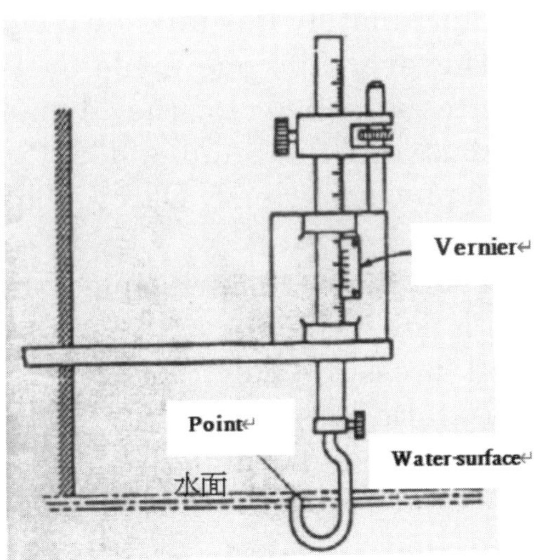

Fig. 4.2 Example of hook gauge [4]

The measurement of the head (h) under the current JIS is like that shown in

Fig.4.2. It is stipulated to read with a hook gauge or float gauge, or with a glass tube with a scale, with an accuracy of 1/150 at the head. In industrial measurement, usually by various levels, it is common to continuously measure with the calculation of flowrate according to the specified flowrate formula.

(2) Pitot Tube (4)(7)

According to a survey by Kesakatsu Koizumi, "The historical development of flow metering" by R. S. Medlock seems to describe the history of flowmeter development in detail, and the first flowmeter was a Pitot tube in 1728. It was first designed as a device to measure the speed of ship, but unfortunately it is unknown what the shape of the device was. However, the one made by Pitot had a funnel-shaped entrance attached to the dynamic pressure hole, that disturbed the flow and did not provide accuracy. It was British physicist Henry Darcy who improved it to the current form.

Figure 4.3 shows the principle of Pitot tube.

When a device having a total pressure hole and a static pressure hole is placed in the flow, total pressure (P_1) and static pressure (P_2) are measured, and the dynamic pressure (P) can be obtained as

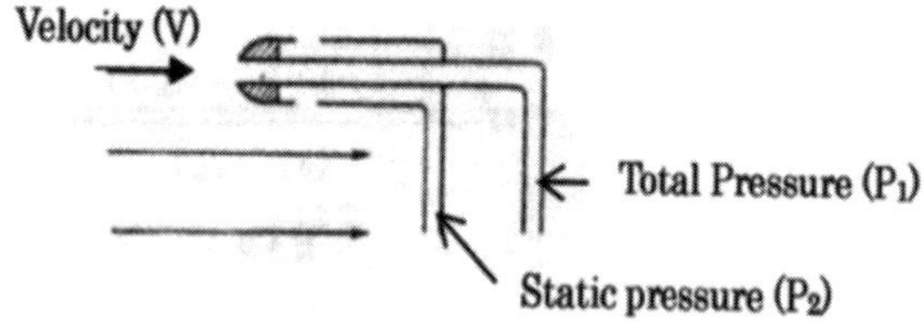

Fig. 4.3 Principle diagram of pitot tube

$P = P_1 - P_2$

and subtract the static pressure from the total pressure to obtain。

From this, the flow velocity V can be obtained by the following equation.

$$V = k\sqrt{\frac{2P}{\rho}} \quad \text{---------------------- (4)}$$

Here, ρ: fluid density, k: correction coefficient.

This system is a Pitot tube.

(3) Lavoisier's Gas Measuring Instrument [3][6]

The balance has a long history from ancient times, as in Fig. 4.4 which shows a steelyard balance found in Pompeii. Therefore, it might have been actually used at least in Roman times. Since then, balances have been used to buy and sell goods, regardless of age, east or west. As mentioned in "2.4.4 To the Era of the French Revolution" the great French chemist Lavoisier has mercilessly dispatched by the guillotine. He was one of those devoted to the international standard of metrology.

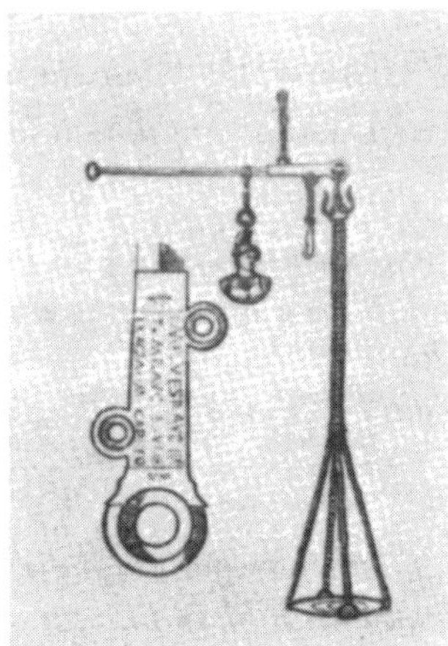

Fig.4.4 Rome balance [3]

At that time, precision balances were actively made by Munier, and Fortin in France, and were designed to weigh 1

Fig.4.5 Lavoisier's gas measuring instrument made by Munier (6)

kg unit with the accuracy of 1 mg. Lavoisier commissioned Munier to manufacture a "gas measuring instrument" as shown in Fig.4.5 in 1783 with the aim of producing hydrogen industrially. The gas density was measured using this gas measuring instrument.

(4) Positive Displacement Flowmeter (5)(6)

Positive displacement flowmeters had existed since the 16th century in Table 4.1, but there is no evidence from literature. Perhaps because the pulley bucket system of the pumping machine has containers. Since it can be a positive displacement flowmeter, it is graphically done so. (Refer to Fig.1.25)

Historically, the first positive displacement flowmeter to appear was the wet gas meter by Clegg in 1815. In the UK, the city gas business broke out at the end of the 18th century, and the supply of gas for coal carbonization lamps began in Glasgow, England in 1792. Samuel Clegg (UK, 1781-1861) had been involved in these gas businesses, contracting and manufacturing gas equipment, wet gas meters were also incorporated into devices in 1815.

However, they were mechanically complicated and did not work by the defect of valve mechanism. In 1819, the wet gas meters were significantly remodeled by Craig's adopted son, John Malam. Figure 4.6 shows this. There was a webbing wheel with one chamber inside and four chambers outside, and when the wheel rotated, gas entered from the inside and was pushed out to the outlet pipe leading to the outside chamber.

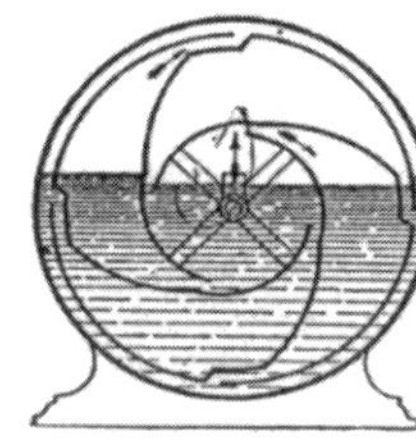

Fig. 4.6 John Malam's wet gas meter (6)

Since the one based on this principle had high accuracy, it had been used as a reference device for gas meter verification. Commercial water meters were designed and manufactured in 1820, five years behind gas meters, because water was probably cheaper than gas. However, 13 patents related to water Meters were registered between 1820 and 1850, and 61 patents were added in just one year from 1850 to 1851, that led to the rapid spread of water meters. Meters appeared

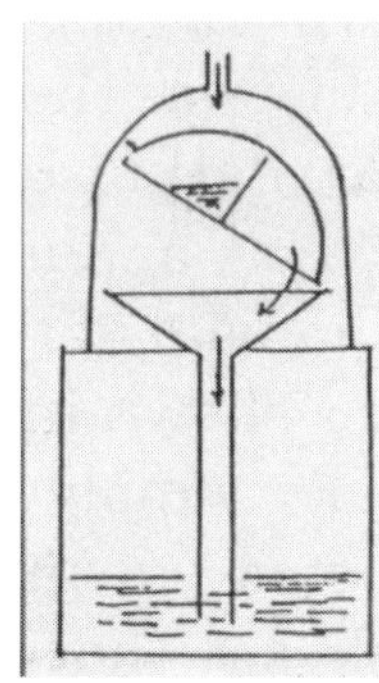

Fig.4.7 Crossley's bucket meter (5)

during this period are roughly classified into the following types:

(a) Measure type or bucket type
(b) Gate type
(c) Piston type
(d) Inference type (turbine and impeller type).

Now, the first water meter designed was probably the bucket meter of Samuel Crossley, made in 1825, as shown in Fig.4.7. It was later redeveloped by Mead. This meter consists of a rotating bucket that is divided into two by a partition wall. When one bucket starts to overflow, the other bucket begins to fill, then a rotational moment is applied, and the full bucket is tilted. It is based on the principle of dropping water into a water tank. The second oldest and best known in the mid-19th century was the one which is a Parkinson's meter that applies the mechanism of the wet gas meter shown in Fig.4.8. A water supply tank is adjusted to a constant level so that water can continuously enter each of four rotating drums. As a result, the drum was designed to rotate freely and discharge water into an aquarium. However, these flow meters had the defect of being unsuitable for pressured water. Flow meters instead of these will be described in the following section.

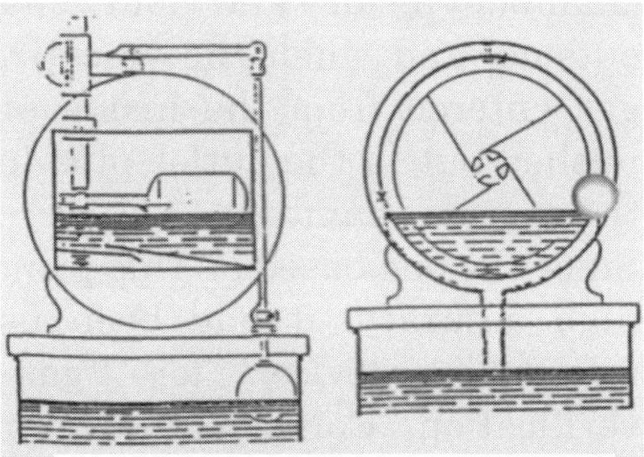

Fig.4.8 Parkinson's meter (5)

4.2 To Flowmeter as an Instrument

4.2.1 History of European Water Meters [8]

H. Bunte wrote the precise review of water meter patents in the specialty magazine "JGW" in 1875 and 1876.

The first high-pressure water meter was the piston-type water meter of Brunton. 146 invention patents of various piston structure type water meters had been registered with the UK Patent Office by 1871, of which itemized 104 from UK, 14 from Germany and 17 from United States and so on.

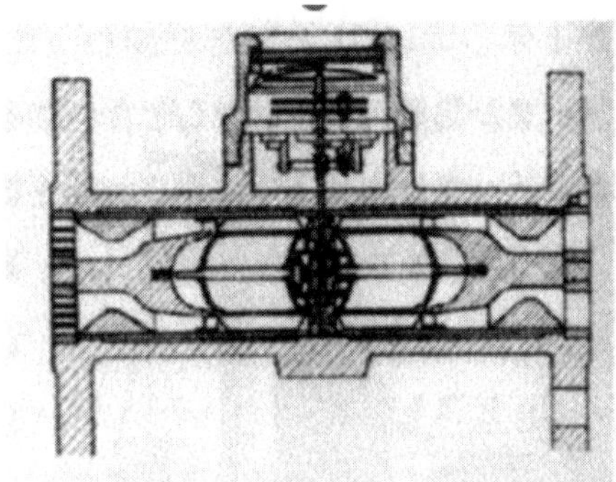

Fig. 4.9 Double cylinder meter by Siemens

First impeller type water meters by Dunn and Taylor separately appeared in 1851 and 1852, and the double cylinder meter as shown in Fig.4.9 was designed by Siemens in 1852. It is worth mentioning that the straightener is already attached to the entrance in this figure. Unfortunately, the shape of the impeller is not really understood.

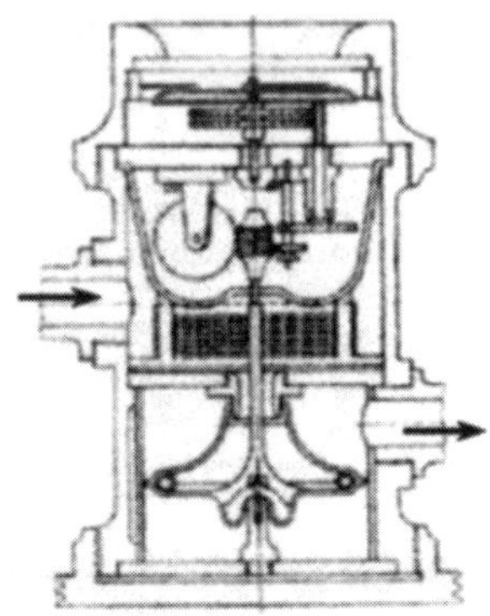

Fig. 4.10 Reaction turbine meter

The reaction turbine meter in Fig. 4.10 by Siemens-Adamson was announced. Water flows in from inlet at the top and flows out from the side of the turbine.

It is probably inspired by the Francis turbine for power generation designed by James B. Francis in 1840.

Siemens introduced the first impeller-type water meter with a dual radiant jet in Figure 4.11 in 1865. The elaborate part of this meter is that parts that seem to be a straightener and filter are integrated into the entrance side.

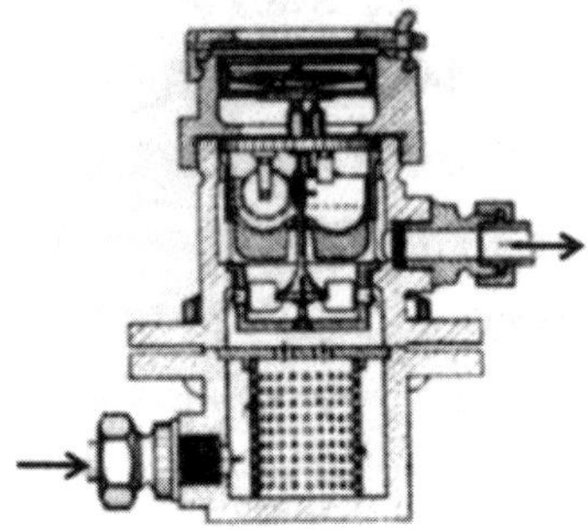

Fig. 4.11 First impeller type water

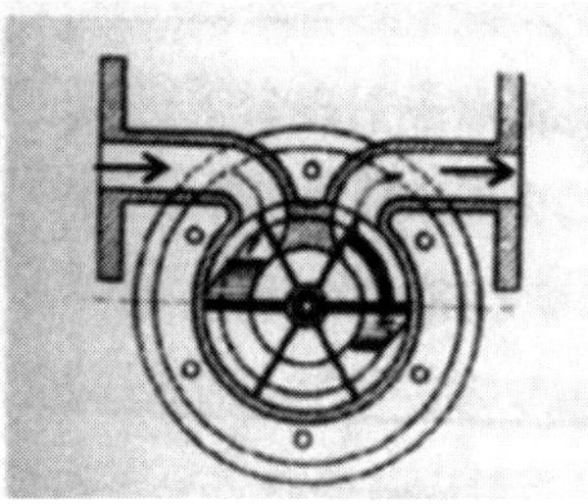

Fig. 4.12 Flat impeller type meter by Rosencrantz

Fig.4.13 Two-rotors meter by Faller

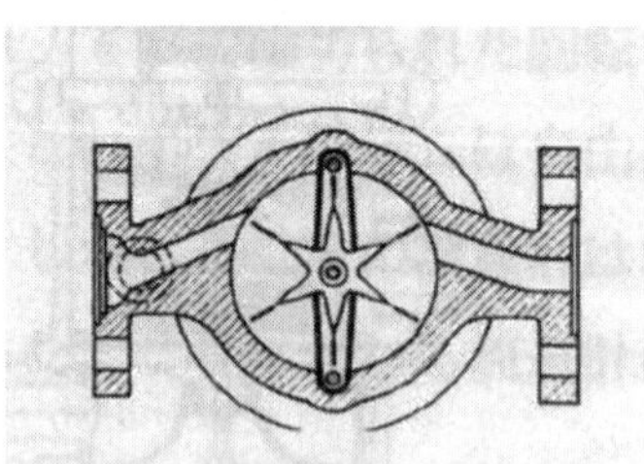

Fig 4.14 Jet meter by Faller

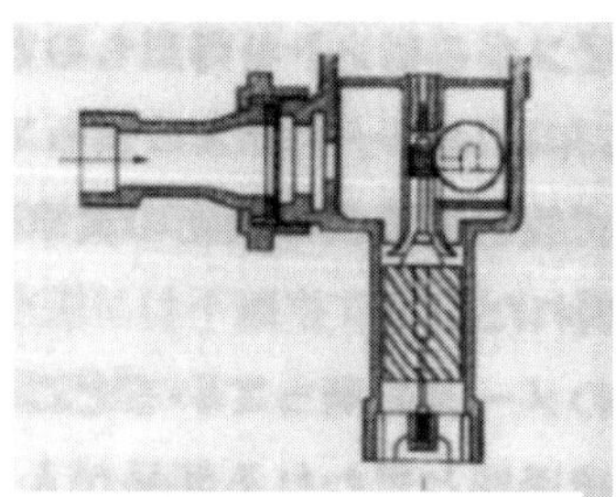

Fig.4.15 Worm impeller meter

Turbine meters, wheel type, worm type, screw type, impeller type, and reaction turbine type, and so on are enumerated. Fig. 4.12 shows a flat impeller type meter by P. Rosenkrantz (Germany), 1871/72. It is consisted of a shaft with six arm-shaped flat impellers.

The two-rotor meter shown in Fig. 4.13 by J. Leopolder (Austria) in 1874, the jet meter in Figure 4.14 by Faller (Austria) in 1875, and the worm impeller meter in Figure 4.15 by G.A. Everett (USA) in 1875 were made one after another.

Figure 4.16 shows the error curves of the oldest water meter group from 1875. 1 is Kennedy's reciprocating piston type water meter, 2 is Rosenkrantz's rotary piston type meter, 3 is Guest & Climbs' reaction turbine type meter, 4 is Faller's multi-jet impeller type meter, 5 is Faller's single jet impeller meter, and 6 is Mainecke's multi-jet impeller meter.

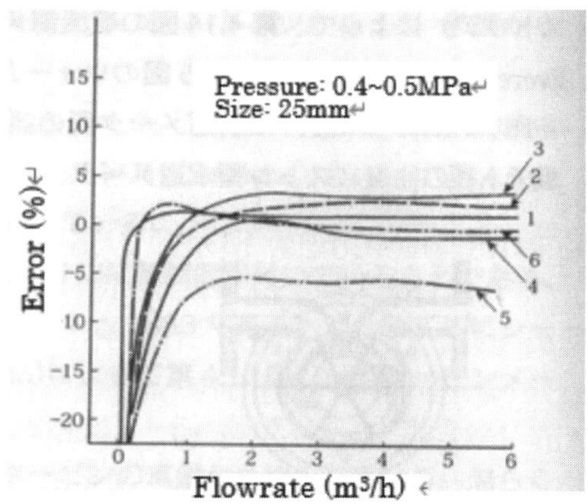

Fig.4.16 Error curve of the oldest water meter

It can be seen that the piston type or multi-jet impeller type meter had wider flowrate range and better accuracy than other meters. In addition to the meters mentioned so far, the Waltman meter had existed for some time.

In the expansion work of Naunhof (a waterworks company in Leipzig), A. Thiem recognized as a hydraulic scholar in 1893 needed a water meter to keep track of water intake. A large impeller type meter could not be used because of the displacement of 30,000 to 35,000 m^3 and high-pressure loss every day and because of location problems. Finally, in 1790, Reinhold Woltman came up with the idea of assembling the measuring impeller like Fig. 4.17 into piping. In this way, the impeller was in the axial direction with respect to the flow and the bore diameter was larger, therefore the pressure loss became many times smaller than with the impeller type flowmeter. The first structure was significantly improved by M. Rother, director of the Leipzig waterworks in 1901. Meinecke mass-produced the Woltman meters in 1896. Meinecke also designed and manufactured a type of disk meter as shown in Fig. 4.18. An excellent (instrumental error) correction device for certification, that was sensitively considered as a measurement technology at the company, was carried as an attachment to the water meter.

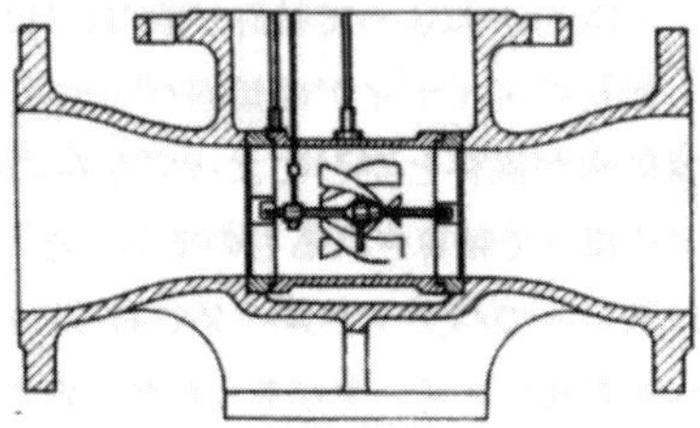

Fig.4.17 Special assembled Woltman meter

Fig. 4.18 Disc meter of Firm Meinecke

In 1898, Siemens Halske also produced something like that.

Rotary piston type meters were produced by Pipersberg Co. and by Bopp & Reuter Co. in 1906, but it is said that they could not compete with the impeller-type water meter.

4.2.2 Water Supply Situation in the United States (10)(11)

German Friedrich Lux gave a detailed speech at the Chicago Conference of ASME (American Society of Mechanical Engineers) in 1893 that the history of water meter development.

"The first patent for water meters is believed to be 1824, with the first piston type patented by Brunton in 1828 and the first impeller type patented by Dunn in 1858.

The first idea was a tilted vessel type, in which two containers were vibrated from side to side depending on whether the container was full or empty, and clockwork was used to integrate the water. (This would be the Crossley's bucket meter shown in Fig. 4.7)

Another was a wet gas meter, but neither could be used under high pressure. A dry gas meter (membrane type) was tried to measure water but failed. The main reason for this was that the water often fluctuated violently and the thin film could not withstand it. Therefore, the one called the space or piston type and the another one called the speed or impeller type that Siemens first made, would be in the limelight.

As a strange idea, gas replacement method was proposed. The idea was to indirectly measure the amount of air which was expelled by water entered in the container, using a gas meter.

The other one was to insert a round bar of alabaster into the discharge pipe, take it out periodically, weigh it, and measure the amount of water by the amount of calcium sulfate dissolved in water. Of course, such things were not allowed to mix in drinking water."

Addition by the writer: By the way, Edison (Thomas Alva Edison, USA, 1847-1931) had a similar idea. That is, in his first electric counter or integrator (instrument) report, "Using zinc as an electrode, the amount of zinc dissolved by the secondary current would be useful in calculating the total amount of fluid that passed through. It was never thought that Edison was interested in measuring water flow.

Photo 4.2 Edison having the idea of water meter

"Since the piston type is firmly incorporated, frictional force is large, therefore it has defects such as large pressure loss, easy wear, poor sensitivity for lesser flow, and high cost. However, the impeller type has the disadvantage that water can pass through when it breaks down, so there is little difference between two types. In fact, the piston type is leading in France, but the impeller type is leading in Germany and Austria. The impeller type is divided into two groups. The first method is the one where water is ejected from multiple ejection holes and hits the blades evenly, so that one-sided wear of the pivot can be avoided. Water meters from Siemens, Meinecke, Valentin and Leopolder etc. belong to this group.

The second method, where all the water spouts from one placc and hits the impeller, distorts one side of the pivot, but it is the most well-known water meter, called the Faller Group, and companies such as Dreyer, Rosencrantz and Droop belonged to this group.

The advantages of the first method are that the spindle wears less, the sensitivity is good, and the accuracy is permanent, but the disadvantages is that the friction loss is large because water is divided into multiple ejection holes, as result, the pressure loss becomes high, and the operating power of water meter increases.

Now, water meters also have two types of indicators, so-called "wet running" and "dry running". The former does not have a stuffing box, so it has less friction and is more sensitive, but the dial quickly becomes cloudy, especially in impure water and water-stopping piping, and if it is made of metal, it will be oxidized or be difficult to see.

In addition, the top of the meter can easily become a stagnant puddle, that can lead to bacterial growth. Third, the glass plate will be easily broken due to sudden increases in pressure.

Static water pressure can be as high as 3, 5 or 8 atm (15 atm needed in Switzerland), and the pressure rises even more when the valve is suddenly closed.

Static water pressure can be as high as 3, 5 or 8 atm (15 atm needed in Switzerland), and the pressure rises even more when the valve is suddenly closed.

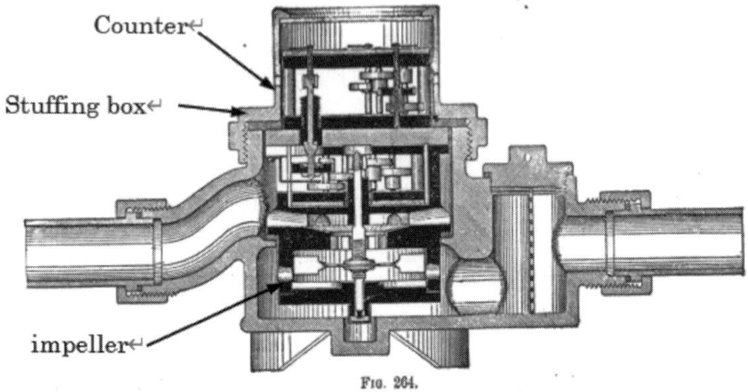

Fig.4.19 Shinzel's ebonite water meter (10)

Then, taking what I said into consideration, I will describe the Shinzel's ebonite water meter that has been developed. This belongs to the Siemens group, and the interior chamber and impeller are made of ebonite. Since the impeller axle is made of delta metal rotated by the ebonite pivot in the polished carnelian bearings, the wear is less. Fig. 4.19 shows the Shinzel's ebonite water meter with a dry running counter, Fig.4.20 shows the shape of the ejection hole, and Fig. 4.21 shows the counter.

Fig. 4.22 shows the results of the durability test in Vienna. After 12 months, the accuracy is getting better and better even after running 5000m^3 (corresponding to 5 to 7 years of normal use)

Fig.4.20 Ejection hole (10)

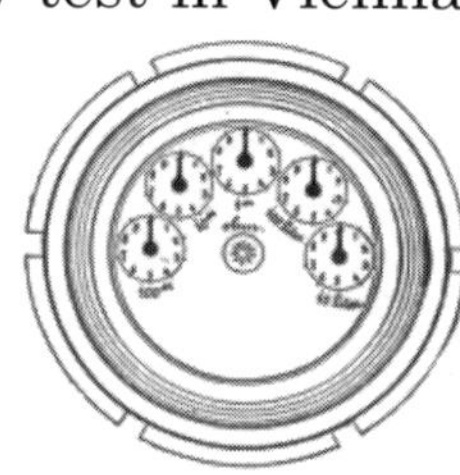

Fig.4.21 Counter (10)

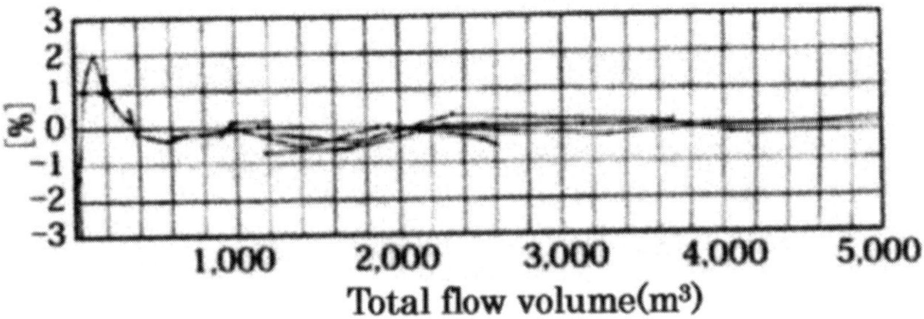

Fig.4.22 Durability test results (10)

and shows a performance that is sufficient for the accuracy of ± 1%.

Here, the introduce of Mr. Friedrich Lux's speech finished. In this way, it can be said that current water meters were almost established 110 years ago. Water meters became widespread in the United States around 1893. It was a long time later that water meters were installed in Japan, but in Japan where water is abundant, accuracy is hardly an issue, and since the purpose was to save water, most of them seems to be cheaper impeller type water meter of German companies to this day. At present, the second group of inexpensive household products occupy the mainstream as a single box type, and the multiple injection type products of the first group are accurate and durable, so they are used for experiments and reference devices.

Figure 4.23 shows a rising disk water meter developed in 1873 by Deacon (G.F. Deacon, 1843–1909). Explaining this structure, the main body is made of cast iron, and the hollow part has a conical inner chamber C made of bronze, in which a disk D is suspended. A thin wire is attached to the tip of the rod of disk, and it is connected to the counterweight W via the pulley R.

When no fluid is flowing, the disk is located in the upper part of the conical chamber.

The disk is pushed down as the flowrate increases. A pen attached to the wire records the movement of the disk on drum E, which spins the mainspring once every 24 hours. Therefore, it was not a water meter that integrates the total amount like a normal household meter, but it was used in large quantities in the UK because it shows the time change of the amount of water.

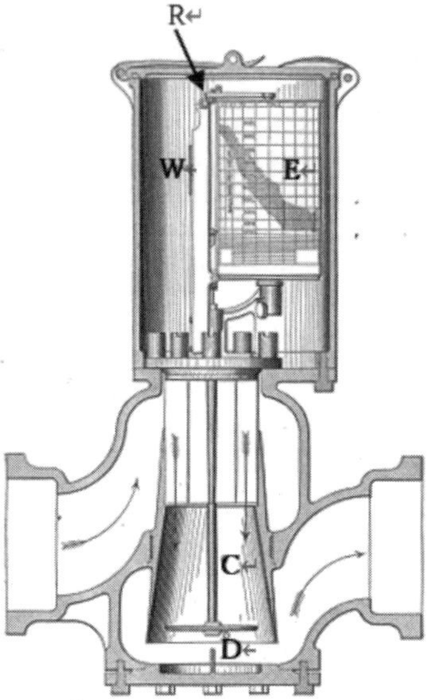

Fig.4.23 Disk water meter by Deacon (11)

Around 1885, the United States was suffering from a rapid increase in water usage. In order to analyze the cause, attention was paid to the Deacon water meter. In 1880, the American Society of Civil Engineers J.P. Davis first installed the meter in Boston. The size was 6 inches and the recording capacity was 25,000 gallons / hour (95m^3 / h), which was about the amount of 3,000 to 6,000 people. Then 83 units were installed throughout the city, which could measure the amount of 420,000 people. This recorded daily trends in water usage, and water usage from 1:00 to 4:00 in the middle of the night, when people should not use water, is about 60% of the average hourly usage per day. Since the cause was the leakage from the water pipe, it was reported that inspectors went around each household for each Deacon and repaired the defective water pipe by relying on the sound of the leakage to rcduce the amount of leaked water.

4.2.3 History of Gas Meters (12)

Regarding gas meters, the wet gas meter which was introduced was developed and improved by Samuel Clegg and John Malam was the (4) positive displacement flow meter of 4.1.1, but after that, Charles W. Hyman of the United States obtained a patent and the Hyman drum was completed.

Photo 4.2 Wet gas meters standing at New Amsterdam Gas Company (12)

The Hyman drum was characterized by a decrease in resistance to rotation in water and an increase in capacitance. Once, it is said that this type of meter was used to measure gas production and delivery at most gas plants.

Photo 4.2 is a typical facility at that time located at the base of the New Amsterdam Gas Company in the Netherlands. It consisted of six large-capacity wet gas meters and had a capacity of 280,000 m^3.

However, aside from the accuracy of wet gas meters, they were notorious for being difficult to use because of temperature change (freezing), and the price was too high. So, Samuel Clegg tried to prototype the first dry gas meter in London.

In 1833, James Bogardus made a meter with an internal unit in a square tin case using a silk diaphragm (membrane) soaked with oil and a valve, but the diaphragm did not work well and the displacement of gas was not constant.

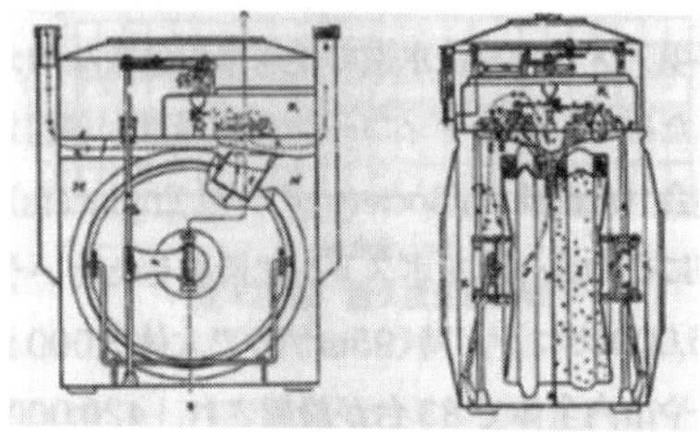

Fig.4.24 Two membrane type wet gas meter by Tomas Glover (12)

After that, William Richard and others proceeded with improvements, and in 1850 Thomas Glover made repeated improvements, and the current two-membrane slide valve meter shown in Fig.4.24 was finally completed.

4.2.4 Venturi Meter (4)(9)

The late appearance of the Venturi Meter (Fig. 4.25) seems surprising, because the theory underlying this instrument has been known from long time ago.

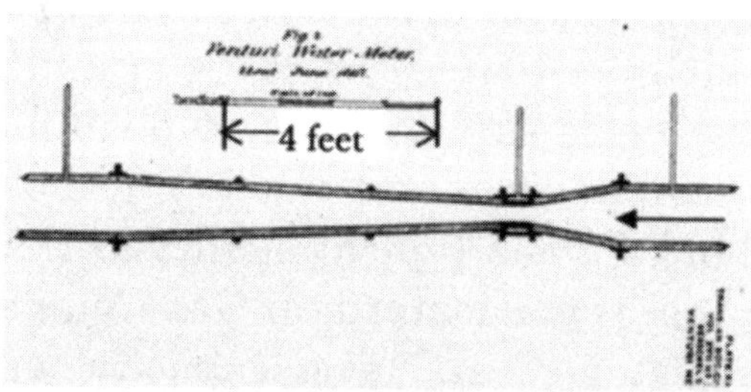

Fig. 4.25 Venturi meter produced as an experiment by Hershel (9)

Already in the Caesar (Caesar, 100-44 BC) era of the Roman Empire, it is said that the principle of restriction by pipe size was used in the distribution of water to Roman households.

At the beginning of the 17th century, Castelli (Benedetto Castelli, Italy) and Torricelli (Evangelista Torricelli, 1608~1647, Italy) developed their respective concepts and derived the basic theory of the restriction flowmeter in that the flowrate is equal to the flow velocity multiplied by the area, and the outflow from the throttle changes depending on the square root of the head (pressure difference).

Photo 4.4 Portrait of Venturi

Bernoulli established "Bernoulli's theorem" in 1783.

Of course, Venturi (Giovanni Battista Venturi, 1746-1822) created the theory of the Venturi tube. Venturi was born and raised in Regio, Italy, in a good environment with a higher education. With a genius brain, he became a bishop at the age of 23 and was then invited to the University of Modena, where he taught geography and philosophy. Thereafter, he was appointed as a national engineer in charge of river and sewerage technology and established a laboratory. In his spare time, he was also studying literature and poetry. It is said that he was likely the return of Leonardo da Vinci. When Venturi was sent to Paris, he was fortunate enough to associate with many scholars and

Photo 4.5 Portrait of Hershel

intellectuals. His name became popular in 1797, "Experimental survey and research on the principle by the lateral transmission of motion in fluid, using an explanation by the differential pressure phenomenon." The Venturi Meter was actually developed by Hershel (Clemens Hershel, USA, 1842-1930).

He repeated many experiments based on Venturi's theory as a water treatment and sewage treatment engineer. He made possible the ability to measure the flowrate of air or water using large-diameter pipes that was previously impossible. After graduating from Harvard University, he studied in Germany and France and became a civil engineer, but he became attracted to hydraulic engineering and devoted himself to measuring the flow of large-diameter pipes from 1887.

Since the orifice had large pressure loss, he made a restriction flow meter with a gently curved diffuser and created the formula of the currently known as a Venturi meter.

However, he named this meter the "Venturi Meter" in honor of the French Venturi experiment. In a world that wants to assume the achievement of the person own thing in many cases, it is truly praised!!

The first Venturi tube made in this year was a large one at 108 inches (2.74m) in diameter and 87 to 89 feet (26.5 to 27.1m) long.

The enforcement right concerning the patent of his Venturi pipe was given to Englishman George Kent and American BIF in 1893 and was commercialized.

Herschel has submitted many reports concerning Venturi Meter to the American Society of Civil Engineers for reference.

As reference (9) is famous, it's abstract will be introduced.

He described how the Venturi meter is superior to the conventional weir-flowmeter, anemometer, and Pitot tube, and manufactured an experimental Venturi meter as shown in Fig.4.25.

From the scale in the figure, it can be seen that the total length is 3.58 m, the bore is 300 mm, and the restricted bore diameter is about 100 mm. This is a 1/9 model of the actual product.

This was tested using the test equipment shown in Fig.4.26.

To explain this device a little, water enters from inlet A of tank B, passes through Venturi and flows out to tank C. Although not visible in this figure, there is a diverter that overflows into tank D.

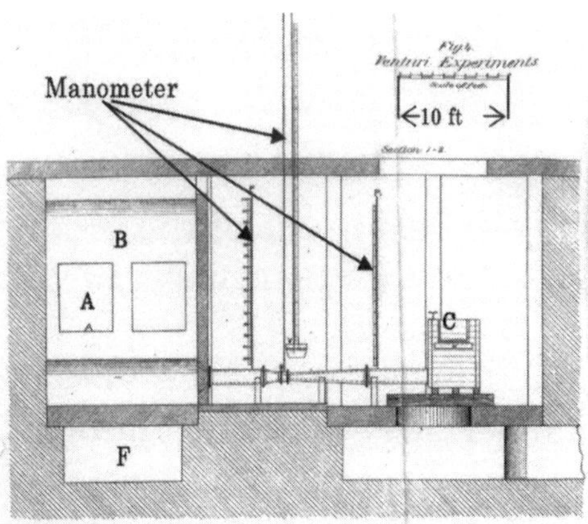

Fig.4.26 Test equipment at Holyoke Water Power Company (9)

Three manometers are installed at the entrance, exit and throat of Venturi, and especially the one at the throat can be measured 34.5 feet (about 10.5m) high enough even in full vacuum.

The test results are shown in Fig. 4.27. The flow coefficient and pressure head are precisely drawn, and the variation of the flow coefficient is within ± 5%.

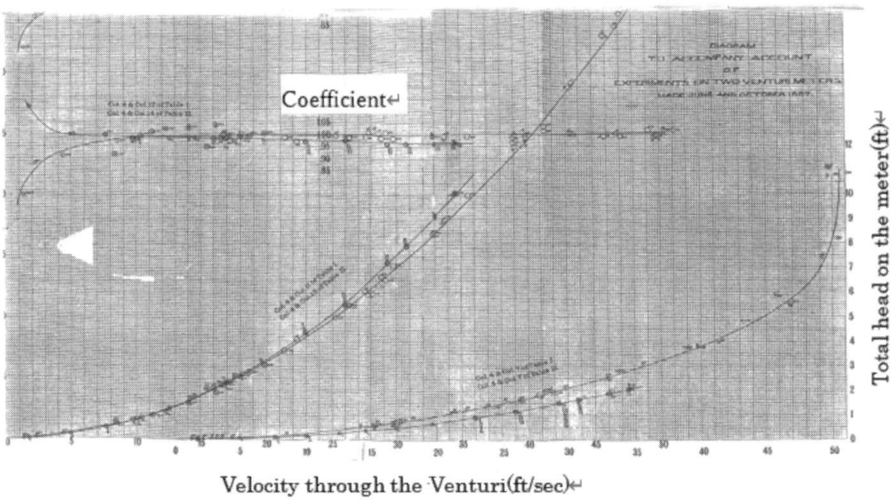

Fig.4.27 Test results of venturi flowmeter (9)

The wonderfulness of the production and experimentation of Herschel's Venturi tube 110 years ago is amazing.

4.2 5. Appearance of the Viscometer

James E. Denton (USA) published a paper on viscometers in the journal of the American Society of Mechanical Engineers in 1887. According to this, over the last decade, oil manufacturers and sellers had been measuring the properties of oils known as "viscosity".

The instrument that measured this property was called a "viscosity meter", which had a simple oil storage tank and

measures the viscosity from the time when a certain amount of oil flows out from a small orifice.
who systematically used it in Standard Oil company products. Figure 4.28 shows an instrument made by G. M. Saybolt (USA),
His measurement of viscosity was quite perfect, and it is said that the same oil sample did not change at all for the entire time. To explain this figure, Vessel A was held in a glass aquarium, in which 4 ounces (113.4 grams) of oil can be placed. At the bottom of this vessel is an outlet O about 1/16 inch (1.6 mm) in diameter, adjusted to be as completely covered by the aquarium as much as possible. A thin glass tube is connected under A. First, the oil level was set at the upper position of a-a, the stopwatch was started at the moment when the oil started to flow out from O at the bottom, and the time measurement is finished at the moment when the oil level reaches the line B. The temperature of the water tank is adjusted by condensed steam. In addition to the Seibolt viscometer, a Davidson viscometer and a Perkins viscometer were also introduced. Of these, G. H. Perkins made a viscometer as shown in Fig. 4.29. It consisted of a cylindrical vessel A in a suitable heating chamber and a piston B, which

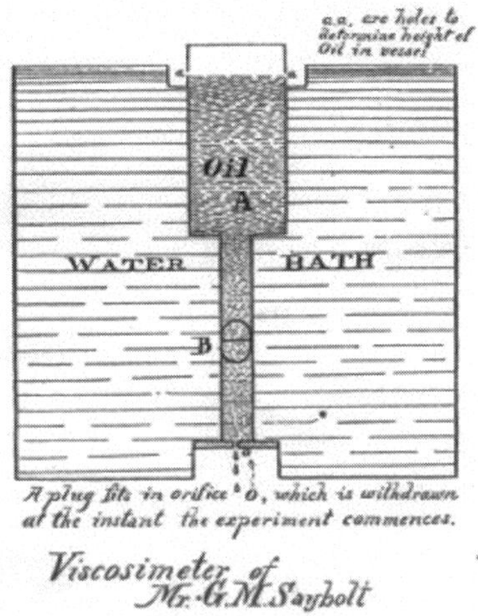

Fig.4.28 Viscometer of Saybolt (13)

was 2/1000 inches (0.05 mm) smaller than the cylinder. The vessel filled with oil and released the piston to descend. a stopwatch was started when point C of the rod passed the upper horizontal line, stopped when it passed the lower horizontal line, and the time was measured. Table 4.2 shows a comparison table of various oil viscosities measured by these three types of viscometers. The

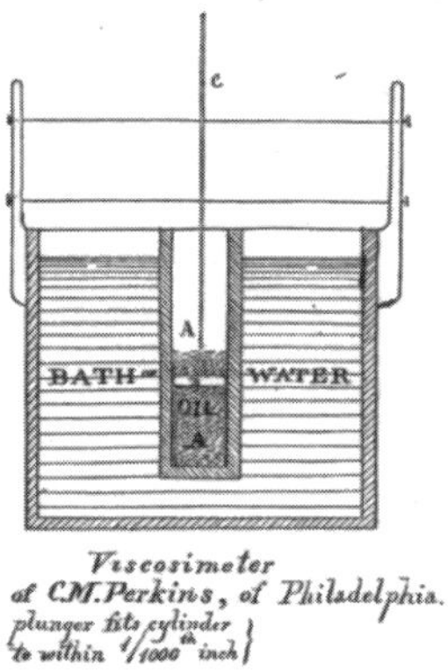

Fig.4.29 Perkins's viscometer (13)

Table 4.2 Viscosity measurement comparison at a temperature of 167 ° F (75 ° C) (13)

DESCRIPTION OF OILS.	VISCOSITY IN SECONDS.			RATIO OF VISCOSITIES.		
	Col. 2.	Col. 3.	Col. 4.	For Col. 2.	For Col. 3.	For Col. 4.
[1] Heavy cylinder oil, pure petro.	346	138	523	1.000	1.000	1.00
[2] Mixed oil {90% " " / 10% Neatsfoot.....	237	73	372	0.68	0.53	0.71
[3] Castor oil	210	71	356	0.60	0.51	0.67
[4] " " {70% Petroleum.... / 30% Animal.......	215	62	261	0.62	0.45	0.50
[5] Mixed oil {33% Light paraffine / 67% Heavy petro..	165	55	....	0.47	0.40	....
[6] Lard	79	42	114	0.23	0.30	0.22
[7] Tallow	....	40	....	....	0.29	....
[8] Sperm	63	35	81	0.18	0.26	0.15
[9] Spindle oil paraffine.........	53	31	53	0.15	0.23	0.10

measurement results for each viscometer of Saybolt, Davidson and Perkins are shown in the columns of Col.2, Col.3 and Col.4 respectively. It was stated that viscosity measurement had become necessary to prevent wear of the bearings of power machines and to grasp the relationship between the amount of lubricating oil supplied and viscosity.

4.2.6 About Newton Again (14)

Speaking of viscosity measurement, "Newton's Law of Viscosity" that was mentioned in 2.4.1, then "Principia" (14) translated by Mr. Masito Nakano was checked again. The title of Volume II "Motion of an Object in Medium that Exerts

Resistance", in which the resistance received by a sphere moving in the medium is defined as follows.

The resistance of a sphere is proportional to the product of the square of velocity, the square of diameter, and the density of medium, if other things are the same.

The law of viscosity is not defined, but there is a sentence as follows:

"It's clear enough from what I've said that the resistance of a fast-moving object is roughly proportional to the density of the fluid moving in it. I'm not saying it exactly. There is no doubt that viscous fluid, even at the same density, will resist more than fluid with higher fluidity, for example, cold oil resists more than warm oil, warmer oil does more than rainwater, and water does more than alcohol, something like that", that is, it stated that if the viscosity of fluid is higher, the moving object in the fluid receives higher resistance.

By the way, this time, it was found out later why Newton was interested in fluid mechanics, especially the resistance that moving objects receive from fluids. It was inevitable. Volume III In the world system,

Proposition 10. Theorem 10. The movement of planets in the sky is being able to survive for a very long time.

"In Volume II, Proposition 40, when a frozen water sphere is moving freely in our atmosphere, the air is drawn in time to draw the length of its radius. It was shown that the resistance of the body lost only 1/4586 of the movement. (An omission of middle part)

In the note of Proposition 22 of Volume II, at the height of 200 miles (about 322 km) above the ground, it was shown that the air was diluted by 1 /75 trillion more than at the surface. Therefore Jupiter, which is revolving in a medium with the same density as the air in the super-sky will not lose one millionth of its motion in one million years by the resistance of the medium." Therefore, the above proposition 10. Theorem 10 could be said. After all, Newton had to inevitably discuss the resistance of objects moving in the medium to prove that the movement of the planets around the sun in the sky and the moon around the earth are invariable. The original book of "Principia" was written in Latin, and it was difficult to read because the proof of the theorem did not use mathematical formulas much and it was proved by geometry, and it is said to be unpopular for modern physics researchers. However, Mr. Masito Nakano explained the important points using modern mathematics in the commentary, so if you are interested, this book is worth a look.

Reference

(1) Mitsutomo Yuasa "Chronology of Science and Culture History"

(2) E. A. Spencer "Flow Standard Past and Present" NEL, translated Japanese by Jyuuji Isikawa

(3) Kesakatu Koizumi "History of metrology" National Institute of Advanced Industrial Science and Technology, Japan

(4) Kesakatu Koizumi "There are many dramas in the development of flowmeters." Oval Specialty No.2

(5) Kesakatu Koizumi "Commercial meter talks about the trajectory to practical use"

(6) Charles Singer "History of Technology 8 Industrial Revolution" Chikuma Shobo

(7) Seizo Matusiro "Revised Flowrate" Korona Inc.

(8) Joachim Böhler "Handbuch der Wassermessung" R. Oldenboung Verlag, München, Wine 1965

(9) *Clemens Herschel "The venture water meter" Transactions of American Society of civil engineers, Vol 17. 1888*
(10) *Friedrich Lux "On water measurements, with special reference to Shinzel's ebonite water meter" Transactions and waste of water of water" Transactions of ASME, Vol.14, 1893*
(11) *Dexter Brackett "Consumption and waste of water" Transactions of American*
of civil engineers, Vol34, 1895
(12) *"50 years to measure and protect" Japan Gas Meter Industry Association*
(13) *James E. Denton "The mechanical significance of viscosity determinations of lubricants" Transactions of ASME, vol.9 1887*
(14) *Isaac Newton's "Principia" translated by Masito Nakano, KOUDANSYA*

5. 20th Century of Great Progress in Science and Technology

5.1 Science and Technology of 20th Century

5.1.1 The Era from Steam to Electricity [3]

The era is just about to move into the 20th century, two world wars broke out, involving the world in an era of disorder and destruction, whereas science and technology made great progress and built the advanced culture of today. Assuming the period until 1945 when World War II ended as one period, Einstein's Special Principle of Relativity in 1905 and Quantum Theory started by Planck in 1900 would significantly change the conventional view of classical dynamics. The Nobel Prize was enacted in 1901, and every year since then, rewards have been given to those who had make outstanding achievements in the fields of physics, chemistry, literature, medicine and peace.

On the technical side, Marconi succeeded in transatlantic wireless communication in 1901, opening the door to the information and communications of today.

In 1903, Wright brothers flew an airplane for the first time, promoting the development of aircraft technology and then shrinking the time space of the world. In the same year, the Ford Company was founded, result in establishing the base for mass production of automobiles in the United States. In 1907, the triode was invented, and along with this remarkable progress, radio broadcasting began in each country in 1923, and in 1925, practical television was invented.

The epoch-making events of science and technology during World War II from 1940 to 1945 would have been the practical application of radar and atomic bombs on the victorious side. However, it cannot be overlooked that the V1 and V2 rockets developed by Germany during the war were also useful for space development in the latter half of the 20th century.

In a word, the characteristics of 20th century technology are “the century of electricity”, assuming the 19th century is “the century of steam", as mentioned earlier.

In the process of establishing a monopoly industry from the 1880s, traditional steam energy and emerging electric energy were working side by side, but in the 20th century, electricity overtakes steam.

Fig.5.1 is a graph showing the total amount of electricity generated in the United States.

5.1.2 Development of the Railway (4)

Let's compare the technological level of Japan with the one of the worlds at the beginning of 20th century. The total electrical generation of United States in1900 is zero in the graph of Fig. 5.1.

However, when looking the journal of the American Society of Mechanical Engineers [4], the equipment used at Boston South Station in 1899 was introduced on 127 pages with photographs, and electrical machinery was already being used abundantly. Photo 5.1 shows the Boston South Station building, and Photo 5.2 shows the marshalling yard with rails of 20 or more lanes. As shown in Photo 5.3, Boston South Station sends steam to engines in a power room from a coal-fired boiler in order to generate electricity.

The engines are operated by steam to generate electricity, and through wiring, all switching points and signals on

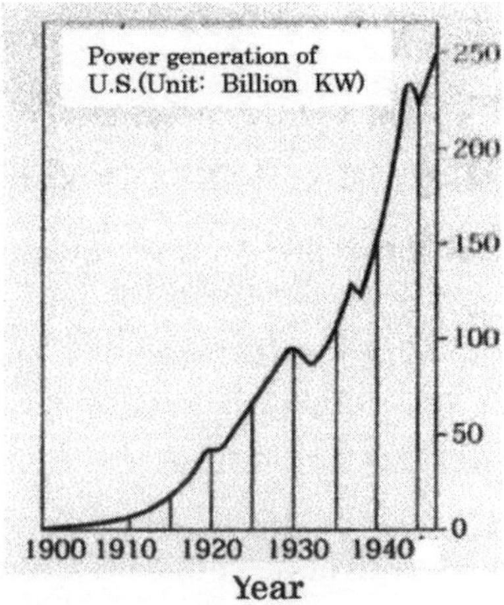

Fig.5.1 Total electric generation of United States[3]

Photo 5.1 South Station Building of Boston[4]

Photo 5.2 Switchyard of Boston South Station (4)

Photo 5.3 Steam engine type generator installed in the power room (4)

operating rails are electropneumatic, and the station building and factory are illuminated using arc lamps and incandescent lamps, and are fully equipped with air conditioning and heating. It had almost reached the current technical level. The pipes of hot water and steam are provided with the same insulation as nowadays.

It is surprising that the technology at that time was quite advanced. It can be said that it is a collection of the best of the technology at that time. Photo.5.4 shows a steam locomotive that was active in the United States. This was published in "Engineering" vol.70 published in 1900 in the United Kingdom, and if compared to

Photo 5.4, 60-Tons Heisler Geared Locomotive, USA

the technical magazines published in the Meiji and Taisho eras in Japan, it had aesthetic photographs and drawings as well as paper quality. There is the huge difference in printing techniques including these

How was Japan in this era?

The first streetlight in Japan was lit in 1882 (Meiji 15) in Ginza, Tokyo.

Fig.5.2 shows people marveling at an arc lamp that was lit for the first time in Ginza. As already introduced on page 81, starting with the opening of the railway between Shinbashi and Yokohama in 1872 (Meiji 5), the introduction of technology had been steadily progressing. The Tokaido Line was opened in 1889 (Meiji 22) and the Tohoku Main Line in 1891.

Photo 5.5 Train crossing the Abekawa Bridge (9)

Fig.5.2 Arc lamp constructed in Tokyo Ginza-Dori 1882 (8)

It is said that Photo 5.5 was a train crossing the Abekawa Iron Bridge taken by Shogun Yoshinobu Tokugawa between 1893 and 1897. (After the Meiji Restoration, he became a marquis and lived near here at that time.) Only eight years later,

Photo 5.5 (B) the First Train in 1809

in 1890 (Meiji 23), an electrical train ran at the Ueno Expo. (See Photo.5.5 (B)). Compared to 30 years after the birth of the first steam locomotive in the UK to the opening of the railway in Japan, unexpectedly shorter than the appearance of the train in Germany to the arrival in Japan, which was 11 years. As a national policy, Japan was able to catch up with and overtake Europe and the United States, and modernization proceeded at a rapid pace. Practical electric railways began in Kyoto in 1893 and the train network expanded in Tokyo in 1904. In this way, photographs and figures showing railway technology from the end of the 19th century to the beginning of the 20th century gave a glimpse of the technological level of the world and Japan.

5.1.3 Tokyo Waterworks (5)

Now, since a valuable report in the 1901 edition of Engineering in the United Kingdom on the construction of the Tokyo waterworks, I will now present an overview.

Water to Tokyo had been carried from various sources by canals and wooden pipes for about 200 years during the Tokugawa dynasty.

A new water supply system was decided upon by the deliberation of the Tokyo city council and all the construction Techniques were proven overseas.

The water supply covered the entire area of Tokyo City, which had a population at that time of about 1.2 million. The amount of water was 4ft^3 (about 0.11m^3) per person per day, with enough to cover 1.5 million people.

The source of water was the Tama River, where water was very clean and suitable for large cities. There were three settling tanks, one of which was 220m long, 104m wide and 6m deep, with a total capacity of about 255,000 cubic meters, which could supply half a day for 1.5 million people.

The water was filtered by sand at a filtration rate of 3m /

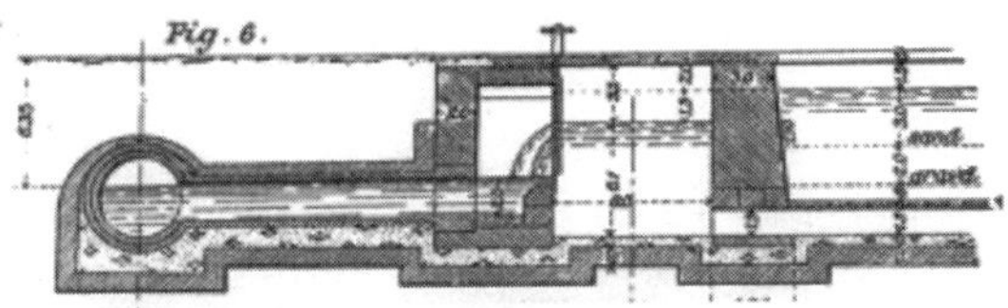

Fig.5.2 Closs view

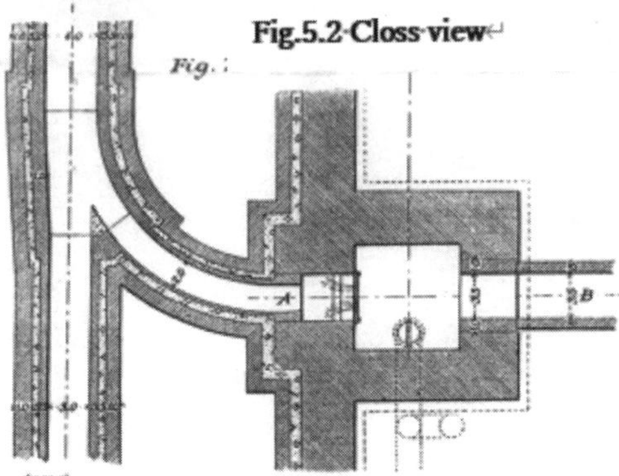

Fig.5.3 Flat plane view

Fig.5.2 and 5.3 Filtration layer map of Tokyo waterworks in 1901[5]

day, and the filtration tank is shown in Photo. 5.5. Reservoirs were set in Yodobashi, Hongo and Shiba. Water was pumped up from the Yodobashi Distributing Station to high places and sent by gravity from a Shiba Reservoir and the Hongo one to Yodobashi pumping station.

Four 306 horsepower pumps were installed, and the capacity of each pump was 1,000,000 ft^3 (about 28,300 m^3) in 24 hours,

Photo 5.5 Pumping up engines manufactured by Kawasaki Heavy Industries [(5)]

and the head was 80 to 100 ft (24 to 30 m). These pumps were made by Kawasaki Heavy Industries Kobe. Six boilers operated these pumps. Various types of cast iron pipes with a diameter of 100 to 1100 mm and a total length of 845,000 meters (about 15,000 tons) were imported from Belgium, the United Kingdom and Scotland.

2,250 large and small sluice valves and 263 consumable water meters were used, the latter being also used for fire protection hydrants.
Most of the valves were procured from overseas, but the consumable water meters, small valves and fire hydrants were made in Japan. The total construction cost was estimated to be 8.5 million yen.
At that time, 20,000 households were using this water, but it would longer for all citizens to draw the waters. Household fittings were brass made in Tokyo. The water charges were 5 yen per year for a family of 5 people, and an additional fee would be charged for each additional 5 people. Water was measured with a water meter, but the meter could be borrowed or owned by an individual, and in the case of borrowing, one with a diameter of 12 mm imposed a fee of 3 yen a year, and one with a diameter of 250 mm it was 75 yen.
For individuals, a test fee of 1 to 15 yen was charged for each person, depending on the size. The metered water charge was 4.5 yen up to $100m^3$ a month, and 4.5 Sen was added for every $1m^3$ above that.
(Author's note: At that time, the starting salaries of an elementary school teacher, university graduate banker, and senior civil servant was 12-yen, 35-yen, and 50-yen, respectively, and a large beer bottle was 19-Sen, (0.19-yen) so it can be said that having a water supply was a considerable expense burden.)

5.1.4 Appearance of Domestic Water Meters (1)

Until then, Japanese water meters had relied exclusively on imports from Germany's Siemens and Bopp und Reuters.

Photo 5.6 Portrait of Daigen Jumonji (1)

Daigen Jumonji (Photo 5.6) was commissioned in 1903 (Meiji 36), by the Ministry of Agriculture and Commerce Japan to investigate windmills and pumping wheels, so he went to the United States. After studying water meters and gas meters in addition to the requested work, he returned to Japan the next year and established Kinmon Shokai (Company).

In 1906, at the recommendation of Shinpei Goto, the mayor of Tokyo, he started research on the production of water meters.

Photo 5.7 The first domestically produced flat disk type water meter

However, the industrial base of Japan at that time was lower than that of Europe and the United States, therefore, there was difficulty with materials and precision machining. However, in 1906 (Meiji 39), he finally succeeded in developing the cross-shaped flat disk type and impeller type water meters shown in Fig. 5.7.

Photo 5.8 Fukuyama type water disassembled

Also, in 1905, Ruizaburo Fukuyama in Osaka City made a prototype of a dry box meter using a floating wing wheel of paulownia painted with lacquer and delivered it to the Waterworks Division of Osaka City the following year. Photo 5.8 is a disassembled one. In Section 5.1.3, introducing an article in the United Kingdom in 1901 about the water supply being laid in Tokyo, but Japan's first modern water supply was completed in 1887 in Yokohama with the design by Englishman H. S. Palma. Tokyo was completed in 1898, 11 years later than that.

5.1.5 Appearance of Domestic Gas Meter [2]

In Japan, gas meters were not produced domestically in the early Meiji era and relied entirely on foreign products, mainly imported from companies such as Parkinson,

Thomas Glover, Manchester, Pradock in England and in the United States.

In 1904 (Meiji 37), Jumonji made a gas meter after great pains with Matsukichi Kuwasawa. Metal-processed parts were hand-processed by tinplate, and dials were paper printing scale attached to a copper plate. It is said that the transmission gear was a handmade art product as follows. First, an original picture was written on paper. Second, it was pasted on a brass plate. Third, the plate was cut with a jigsaw, and finished with rasp. Photo 5.9 is a so called Jumonji's dry gas meter A type. Thus, water meters and gas meters have a long history. In particular, it is not difficult to imagine that some of the various meters developed as water meters would gradually develop into positive displacement flow meters and turbine meters for industries such as the petroleum and petrochemical industries.

Photo 5.9 Domestic product Jumonji's dry gas meter A type : Left; outward, right; perspective [2]

5.1.6. Invention of Airplane and Birth of Modern Fluid Dynamics [6][7]

In the early 20th century, steam locomotives and electrical trains were put into practical use as mentioned above. Fluid dynamics came into the limelight as Wright brothers first flew an airplane in 1903 and became more interested in aircraft.

Ludwig Prandtl (Germany, 1875-1953) presented the "Boundary Layer Theory" on fluid motion at the International Mathematics Conference in 1904, consisting of only eight pages. This was highly evaluated and he was invited as a professor of applied dynamics at the University of Göttingen. This "boundary layer theory" had made great strides in fluid dynamics and would make a large contribution to the subsequent development of aircraft. Since then, he is said to be the founder of modern fluid dynamics for his achievements in supersonic flow, shock wave theory, and wing theory.

Photo 5.9 Portrait of Prandtl (6)

Under the professorship of Prandtl, a non-member professor Kalman (Theodore von Karman, Germany, 1881-1963) was conducting a long-axis buckling experiment, and his colleague Hemens measured the resistance of the cylinder against the flow velocity by using a water tank at the command of Prandtl. Vortices were generated alternately behind the cylinder, and the cylinder vibrated back and forth in the flow, as a result, he couldn't measure precisely. Professor Prandtl said, "The roundness of the cylinder is probably bad," and he had

Photo5.10 Portrait of Theodore von Karman (6)

a hard time polishing it. Karman saw this and thought that the Hemens experiment might not be bad, but that the vibration caused by the vortex streets might be a natural appearance. Therefore, he conducted a theoretical analysis of the arrangement of vortices on weekends. The vortices of the symmetrical arrangement as shown in Fig.5.4 (b) were unstable, but in case of the phenomenon of (a) witnessed in the Hemens experiment, the arrangement was asymmetric (c) and stable.

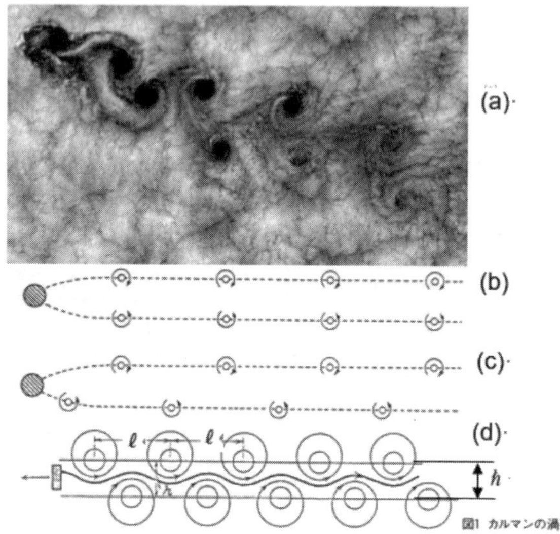

Fig.5.4 Karman vortices street [7]

In this array of vortices, when the ratio of vortex spacing h to pitch ℓ, is shown in the figure (d),

$$h/\ell = (\mathrm{arccosh}\sqrt{2})/\pi \quad \text{------------------------------}(5)$$

It was clarified that the array was stable.
If this phenomenon is expressed by frequency of vortices f (Hz),
The following relation holds. Where, v is the flow velocity.

$$f = v/\ell = 0.281\,v/h \quad \text{------------------------------}(6)$$

Therefore, he derived an important formula that the frequency is proportional to the flow velocity v and inversely proportional to the distance h between top and bottom vortices. The following Monday morning, when he showed the results to Hemens and Prandtl, Prandtl solemnly said, "I'll introduce you to the Academy of Göttingen, so write a paper right away." This paper in 1911 was a successful treatise that remained in posterity, after which such a vortex street became known as the Karman vortex street. A street named "Karman Street" was created in Berlin to celebrate Karman's achievements.

This Karman vortex street is an obstacle to the flight of aircraft, and how to weaken it leads to performance improvement. However, this Karman vortex street is the principle of the Karman vortex flowmeter developed in the latter half of the 20th century, and the key is how to strengthen it here.

5.2 Measurement of Natural Gas at the Beginning of the 20th Century

5.2.1 Flowmeter at the Beginning of the 20th Century

As mentioned above, impeller type flowmeters were already used for so-called domestic flowmeters for water and gas, but industrial flowmeters used in the early 20th century were Weirs, Pitot tubes, Venturis and Orifices. During this time, there were many technical reports concerning their usage and test results, especially natural gas flow measurements in the United States.

(1) Research on Rotary Pumps [10]

Under such circumstances, a report on "rotary pump" was made in 1902, and Wilkin (John T. Wilkin) made the following Summary report.

"The literature on rotary pumps is detailed in the "Kinematics of Machinery" written by Professor Reuleaux in 1876, but it is difficult to obtain. According to him, four-leaf pumps or a four-toothed two-blade pumps had been already in use since 1630.The development and commercialization of rotary pumps was in step with other types from 1860s to 1870s, but in order to counter the price of centrifugal pumps, the blade contact surfaces were made with the cast. Consequently, the leakage under high pressure was large and the performance was very bad, so it became completely unpopular. Therefore, after careful research for about 10 years, a rotary pump with two identically shaped blades was manufactured. These rotors were machined into a complete upper and lower cycloid as shown in Fig. 5.8. If properly assembled and provided, it gives outstanding results. The discharge per revolution of a two-leaf cycloid pump is exactly the volume of a cylinder in which the blade rotated.

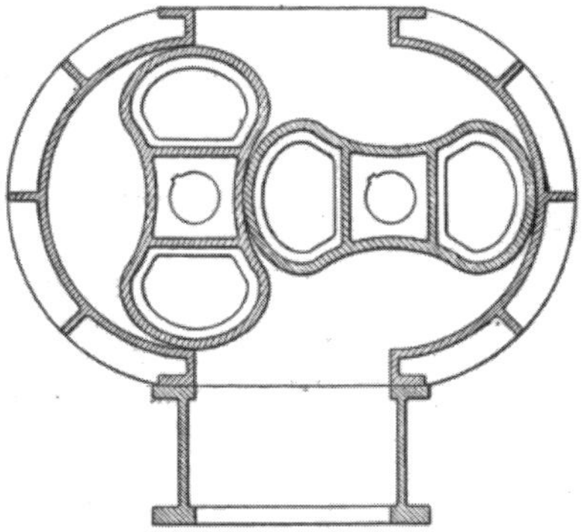

Fig.5.5 Two-blade rotary pump [10]

In other words, the area of the two blade is a half of the

circumscribed circle's area. This fact allows this form to be used as a meter (flowmeter) and its discharge can be measured very accurately."

It suggests a flowmeter now called a roots-meter. This will be described later, when the roots-meter was marketed.

(2) Pitot Tube (11)

In 1903, WB Gregory elaborated on the "Pitot tube" with Fig.5.6 which is a drawing sketched by Pitot and Fig.5.10 which is the improved version by Darcy and Basin. It seems that the figure is small and difficult to see, but if you look closely, the tip of glass tube is closed for Pitot's, and the tip can be opened for Darcy's. If the tip is closed, when the pressure rises, the air trapped in the tip is compressed and becomes equal to the water pressure, so the water column should be lowered accordingly. It

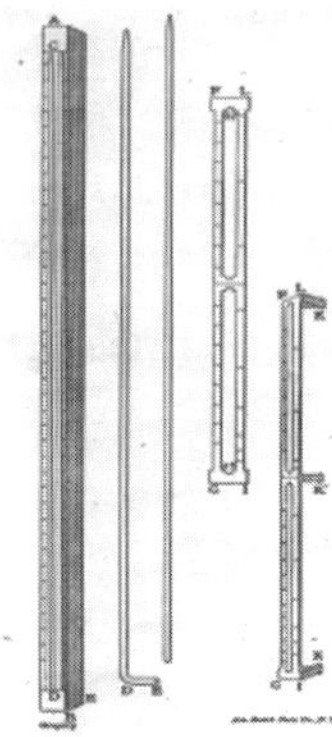

Fig.5.6 Pitot Tube by Pitot (11)

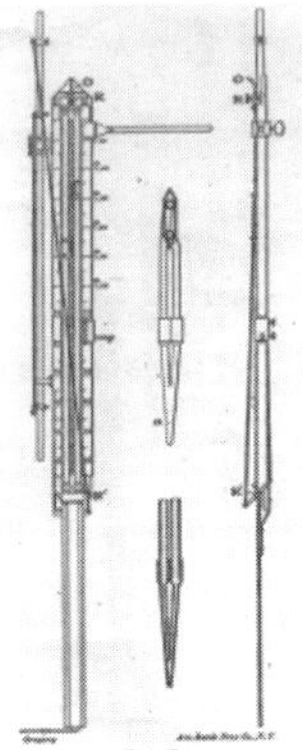

Fig.5.7 Pitot Tube by Darcy & Basin (11)

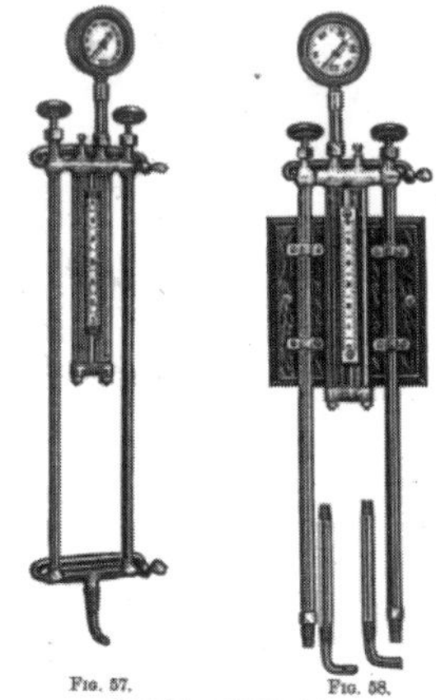

Photo 5.11 Pitot tube patented in the U.S.A in 1892 (11)

seems to be one of the
reasons why the error in Pitot's was large.
Photo 5.11 is an example of a Pitot tube used around 1902. This improved on the defect in Pitot's one by using an equalizer.

(3) Flow Measurement of Natural Gas

The measurement of natural gas was detailed in a report (13) by T. R. Weymouth in 1912. As well known, there were so many natural gas wells in the United States, but the most important issue encountered in natural gas technology was buying and selling gas how to accurately measure large volumes of gas when buying and selling gas.

The problem of measuring natural gas can be divided into two types: open flow and closed flow. There were two types of open flow measurement at that time.

1) The method using Boyle-Shar's law.
2) The method by measuring flow velocity with a Pitot tube.

For 1): A tank is placed into a gas well, the natural gas stored in the tank is released into the atmosphere to make the inside pressure equal to the atmospheric pressure (P_0),
then the release valve is closed and the inlet valve opened. So, the natural gas flow in.
The pressure (P_m) is red.
Assuming that the volume of a tank is V, the amount of inflow (Q) is;

$$Q = V\{(P_m + P_0) / P_0 - 1\} \text{ ---------------------------(7)}$$

For 2): This is a method of measuring the flow velocity with above-mentioned Pitot tube and converting it into a volume.

Next, candidates for flow measurement in the pipe flow were Venturi meters, orifices, Pitot tubes, and Thomas electric meters.

Regarding the Thomas meter here, the diagram of principle is based on Fig. 5.8

Thomas (C.C. Thomas) developed this flowmeter in 1911, but it was a sign of the thermal mass flowmeter that is currently widely used in the semiconductor industry. This principle is to heat all the fluid with the heater in the center, and when the temperature difference (Δt) before and after that is measured, the electric power (E) required for heating is expressed by). Equation (8)

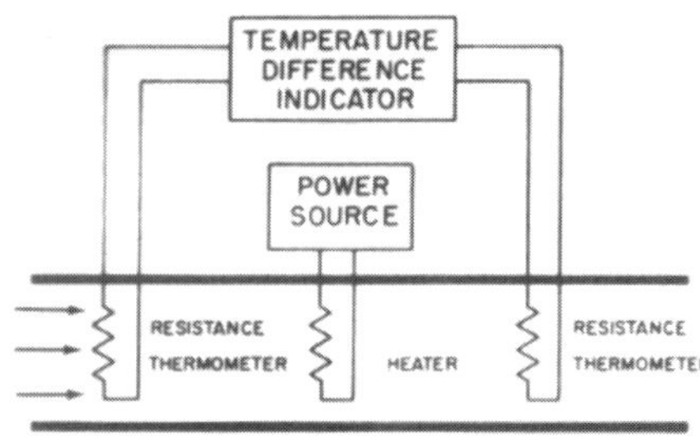

Fig.5.8 Tomas flowmeter

$$E \propto M \cdot Cp \cdot \Delta t \text{-----------------------------------} (8)$$

Where M: mass flow rate

Cp: Constant pressure specific heat of gas

The Thomas flowmeter was a very accurate instrument and was relatively simple to configure, but it required about 1 kw・h of electrical energy to measure the amount of natural gas of 75,000 ft^3 (2,124 m^3). This fact was a major bottleneck, especially in the use of on-site natural gas

Concerning Pitot tube method, Weymouth also points out a Drawback in the flow velocity distribution changes depending on the surface roughness of pipe because the Pitot tube measure the flow velocity of a part of the in-pipe flow and converts it to the total flow velocity.
In conclusion, he also recommended an orifice with a simple structure. As Fisher (Francis P. Fisher) stated in the 1916s, the demand for natural gas had suddenly increased in the last two decades, supplying gases for 2.5 million people. Metering was required for this.
For a few years, bypass flowmeters with1/50 to 1/100 split-flow ratio had been used, but dirt was a problem. Pitot tube and Thomas meter were considered, but in the end, he recommended orifice for large-capacity natural gas measurement.

Photo 5.12 Orifice meter installed in the well (11)

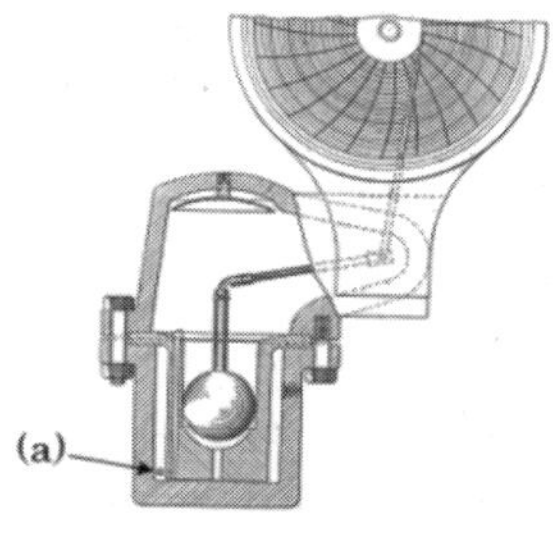

Fig. 5.9 Differential gauge of Ballard (11)

Photo 5.12 is an orifice meter installed at a site, which is a landscape that looks like the site. It is also necessary to automatically record the flow rate, and the differential pressure gauge shown in Fig. 5.9 was developed. This was recommended by H.O. Ballard.

This gauge was designed so that when the gas pressure exceeded the gauge range, the mercury level in the manometer dropped below the opening (a) in the figure and did not damage the mercury body. When the pressure was down, it was ingenious that the mercury returned to its original position. It is a so-called pen recorder in which the vertical movement of the ball is transmitted to the pen.

Now, while measuring the flowrate of natural gas in this way, it was realized that natural gas did not conform to Boyle's law, and Robert F. Earhart said as follows, "Since the end of the 19th century, it has been pointed out that Boyle's law does not fit with certain gases, but it is probably Professor Walker in 1912 who said it first that natural gas does not fit Boyle's law, and in 1914 reported to ASME again under the theme of "Physical Law of Methane Gas". The test was conducted at the Physics Institute of Ohio State University using the equipment shown in Photo 5.13 and Fig.5.10 In this figure, natural gas filled a calibrated glass tube with an inner diameter of 3 mm and set

Photo 5.13 Test device [14]

in a pot containing mercury, and the pressure was changed by a weight-type pressure generator (B).

The height of the mercury column at this time was read on a scale with a magnifying glass (F), and the change in the volume of the contained gas was measured.

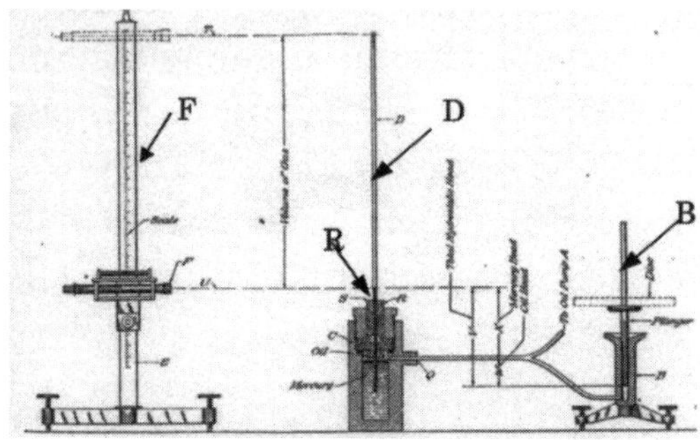

Fig.5.10 Detailed Drawing of Test device

As a result, Fig.5,11 shows a part of the result of measuring the deviation from Boyle's law in this way.

From this figure, the volume of natural gas at 49 atm is about 16% smaller than that of Boyle's law. (Note: In this figure, it is difficult to see the difference between the PV line of Boyle's law and the actual PV line of the law, so refer to the deviation curve.)

Currently, the gas state equation, which is obtained by multiplying Boyle's law equation by the deviation coefficient Z, is used.

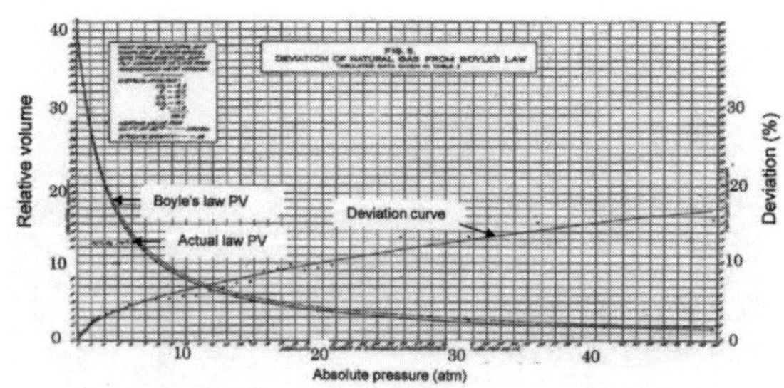

Fig.5.11 Deviation from Boyle's law of natural gas (14)

$$PV = ZRT \quad \text{------------------------------------} \quad (9)$$

Where,

P: the absolute pressure of gas,
V: the volume of the gas
T: the absolute temperature of gas
R: the gas constant.

Since the deviation coefficient Z is the functions of each component ratio, the critical pressure, the critical temperature, the process temperature, and the pressure of the mixed gas, it is complicated.

(4) ASME Pressure Fitting Standard

Looking at ASME at this time, there are reports of water meters and meters for natural gas, but the reports of meters used in oil plants could not be found out.

Since the Ford Motor Company was founded in 1903 and started mass production of cars in 1911, the problem of flow measurement at oil plants and gas stations might be a little later.

However, in the United States, since the piping from oil wells and natural gas wells was mandatory, the American

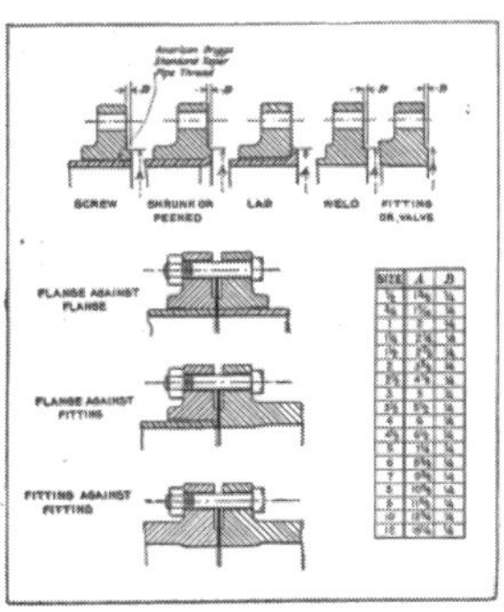

Fig.5.12 Flange diagram for pipe connection (15)

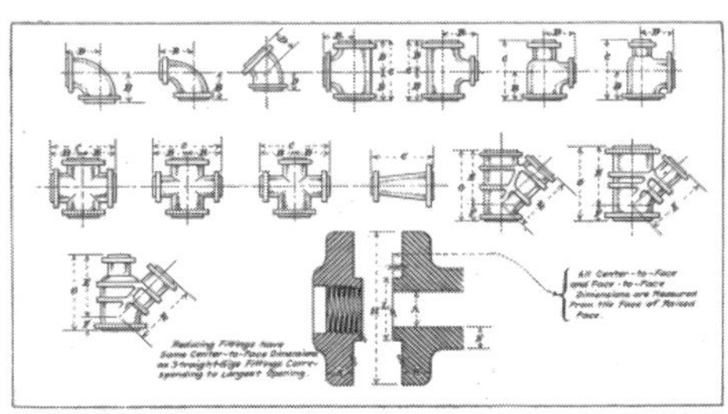

Fig.5.13 Piping fittings diagram (15)

standard for flanges and pipe fittings had been enacted in 1914. Fig.5.12 shows the flange standard and Fig. 5.13 shows pipe fittings at this time, but it seems that there is no big difference compared to the current standard.

By the way, 1916 was the era of the 5th year of the Taisho period in Japan. What about the technical situation was in this era? For example, Kotaro Honda discovered KS magnet steel in 1916, and the Institute for Materials Research was established at Tohoku University in 1918, so it can be said that the spread of science and technology had finally begun. Looking at when technical magazines were first published, "Electrical World" was in 1908, "Electrical Criticism" was in 1913, "Steel" was in 1915, "Mechanical Society Journal" and "Hitachi Criticism" were in 1918. Looking at Japan fluid measurement technology, it was an era where no article about flow measurement appeared in technical magazines.

5.3 History of Differential Pressure Type Flowmeter

5.3.1. Rapid Development of Differential Pressure Type Flowmeter (16) (17) (18)

Various differential pressure type flowmeters were rapidly developed in the 20th century (see Fig. 5.14). However, in the early stage, since indication / recorder or integrator with sufficient accuracy did not exist, very good evaluation was not obtained.

However, on the contrary, it promoted the development of pressure-mechanical conversion devices.

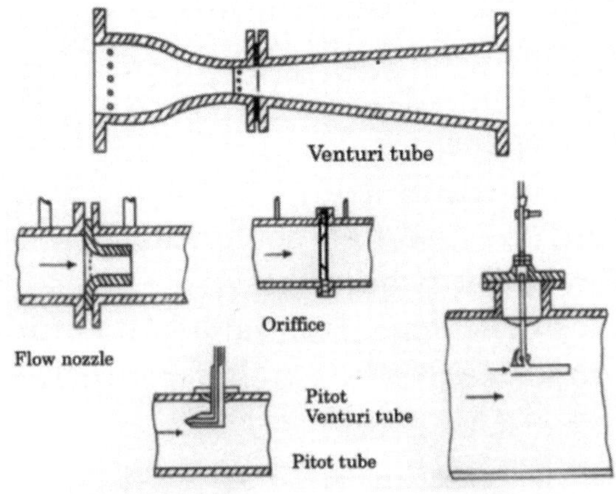

Fig.5.14 Differential pressure type flowmeter quoted from Japanese book in 1925 (16)

Regarding the direction of development of flowmeters in the first half of the 20th century, R.S. Medlock listed the following items.

1) Development of receiving instrumentation
2) Development of square edge orifice
3) Improvement of basic fluid equipment
4) Development of fluid utilization technology
5) Advances in fluid dynamics theory
6) Standardization

(1) Development of Receiving Instrumentation

At the end of the 19th century, when the Venturi tube type had been widely used, the differential pressure measuring device that was the basic principle of the Venturi tube type had been improved in various ways. However, in its early days, it was mostly a U-shaped mercury column and used a large spindle and a cast iron float to obtain the output for driving the recording device (See Fig.5.9 of 5.2.1). When the differential pressure was small, the water column pipe type was used.

When gas flow rates could be measured, a bell-shaped pressure balance using oil seals was also developed. A few years later, a Ring balance with the same differential principle was developed. It is called an annular balance type pressure gauge. (According to Reference 18, the principle is probably based on Fig.5.15 and Fig.5. 16). But the big problem that had to be solved was the technology to derive the square root. Various techniques had been used, such as incorporating cams and linkages into indicator or recording devices. For the example of an integrator, the mechanism to keep linearity was a problem, and in order to drive a device that combines a complicated cam and link mechanism, a powerful and accurate clock device was required, and as shown in Photo 5.14, surprisingly a weight of 60 kg was used. The demand for high accuracy had produced various

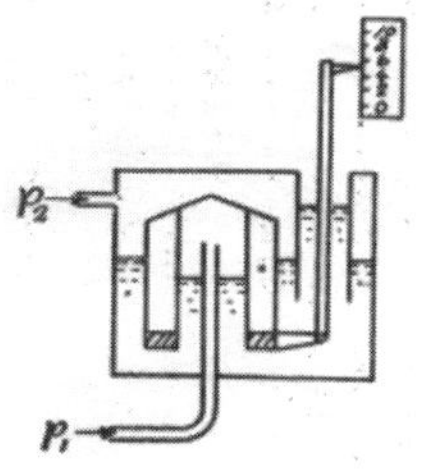

Fig.5.15 Sinking bell- shaped pressure gauge (18)

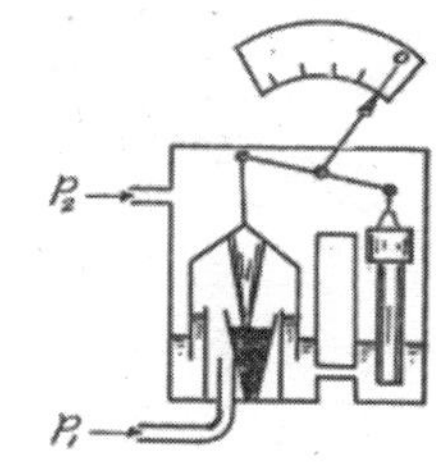

Fig.5.16 Circular Link balance type pressure gauge (18)

Photo.5.14 Early integrator (19)

original inventions, but in the end, it became a mechanical calculator that could operate, do addition / subtraction, multiplication / division and the square route. But these assemblies required complicated and skillful technologies. In addition, since a large amount of mercury was used, the problem of mercury pollution had arisen, but this problem was finally solved by the development of a dry flowmeter with the invention of the diaphragm and bellows in 1907.

(2) Transmission and Recording [16]

When using Venturi pipes in water supply facilities, signals had to be sent from a pumping station or reservoir to a control room, and around 1910, potentiometer-type or resistance-type transmitters were put into practical use. One of the interesting developments of the 1920s was a voice transmitter, that puts the voice of a person reading the position of float into a recording board with a sound tube and sends the voice by phone. It can be said that the first half of 1900 prepared the foundation for accepting electronic technology that would come in the era when such mechanical receiver instrumentation technology steadily advanced.

(3) Development of Orifice Type [16]

It was the middle of the 17th century that the principle of the orifice began to be understood, and it seems that the first patent was granted to Max Gale. It had rounded edges on the both sides of upstream and downstream, and the amount of fluid passing through was calculated by a formula derived from Navier around 1850.

However, since this no sharp edge orifice was accurately unsatisfactory, Gale painfully improved it and in 1909 developed a practical steam flowmeter known as the Sarcometer. It is said to be his greatest achievement.

Next, in an experiment by Thomas R. Weymouth, used a plate-shaped orifice with sharp square edge to measure large volumes of natural gas, with pressure taps at 1 inch both upstream and downstream of the orifice. It later became the American standard.

Also, refer to 5.2.1(3) about the Weymouse.

In 1913, E. O. Hickstein published data on applying a pressure taps to a position 2.5 times of pipe diameter on the upstream side and 8 times of one on the downstream side.

In 1916, Horace Judd proposed the application of the Vena Contractor Tap, for the first time referring to an eccentric, notched orifice.

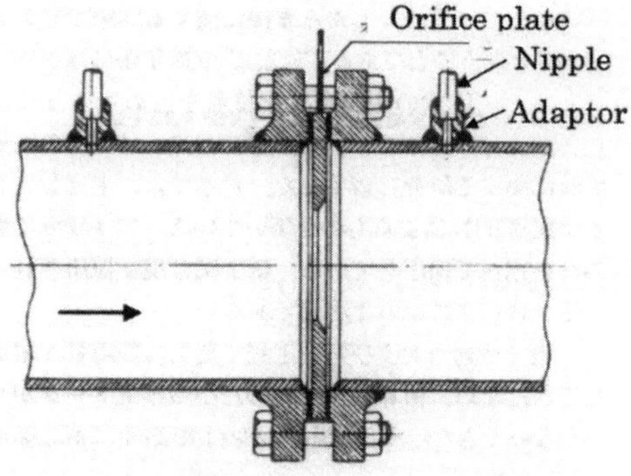

Fig.5.17 Current D, D/2 taps orifice (21)

The flow coefficient data for the Vena Contractor Tap was completed and published by ASME in 1935.

Meanwhile, in the UK, Hodgson was inspired by Gale's work to develop a P-shaped square-rimmed thin plate orifice with a D, D / 2 combination taps as used today in 1909. This orifice had a 45-degree chamfer on the downstream side. Fig.5.17 shows the current D, D / 2 tap orifices.

Hodgson also developed many orifices between 1909 and 1924, setting flow coefficient curves for various aperture ratios and a wide range of Reynolds numbers. Under these circumstances, in Germany, the pressure tap was used as an orifice. In Germany, two corner taps were developed and used that could be taken out directly from the upstream and downstream surfaces. Fig.5.18 shows the corner taps currently in use.

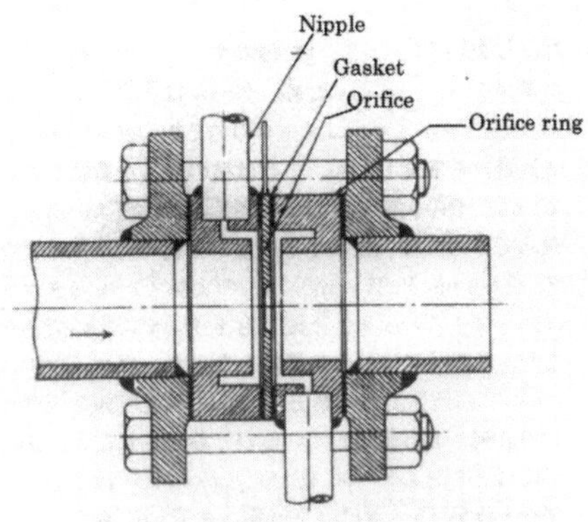

Fig.5.18 Current Corner Tap Orifice

(4) Development of Basic Fluid Equipment (16)

When Herschel developed the Venturi tube, he seemed to think that a long-tapered tube was not needed on the upstream side. Hodgson also demonstrated this in the product that was clarified in the "Olivent" designed in 1912, and contributed to miniaturization and cost reduction.

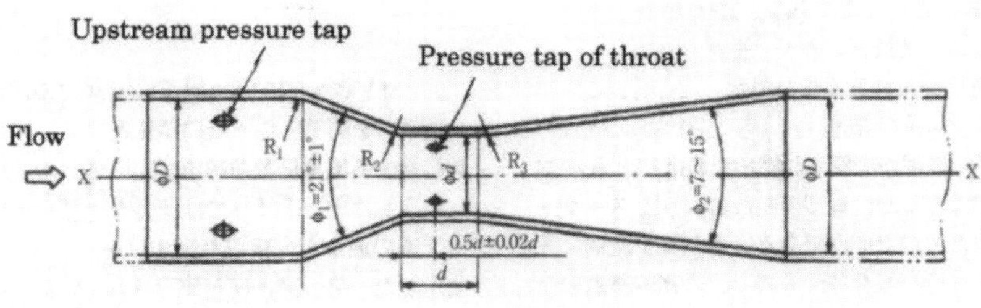

Fig.5.19 Current Conical Venturi Tube (17)

Fig. 5.19 shows the current standard Conical Venturi Tube. The nozzle of flow measurement dates back to Frode in 1847. It is said that he was inspired by an experiment on the friction of a plate pulled in water.

Around 1928, the flow nozzle became a major part of the differential pressure type, and in 1930 Germany began research on the flow coefficient of short flow nozzles. In Japan, the one based on the German standard DIN was mainly used. This was also adopted by ISA in the United States. Fig.5.20 is a dimensional diagram of the flow nozzle currently used (ISA 1932 Nozzle).

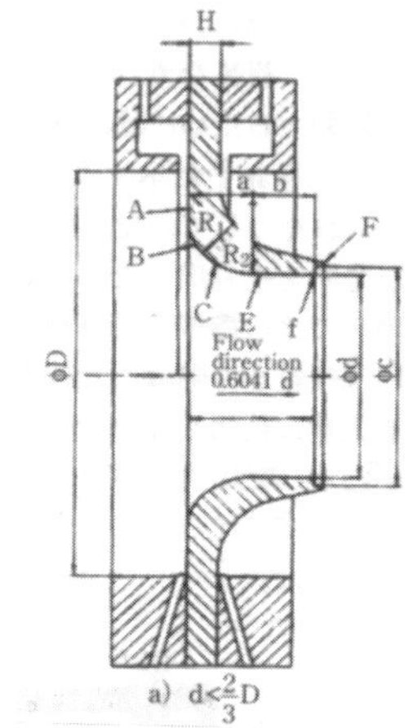

Fig.5.20 Current Flow Nozzle (21)

As the amount of steam used in power and processing processes increased in factories, a small steam meter including an integrator became necessary. So, in 1923, Hodgson developed a shunt type meter that used an orifice for the main stream and an impeller for bypass as shown in Fig. 5.21.

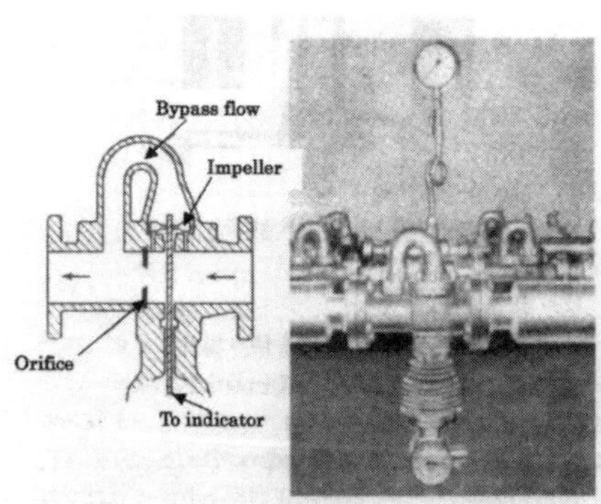

Fig.5.21Current shunt meter developed by Hodgson (22)

The principle of this meter is that the differential pressure before and after the orifice causes a flow in the bypass line and rotates an impeller. The rotation of the impeller is taken out, and the amount of steam is measured by the counter as shown in the photo on the right. Another problem that embarrassed engineers and researchers throughout this period was the measurement of pulsation flow.

Hodgson also conducted thorough research on this and introduced what is called the Hodgson number. The Hodgson number was conceived to estimate the size of buffer tank required to attenuate the pulsation of flow measurement system.

5.4 Progress of Differential Pressure Type Flowmeter and Measurement Mechanism (21)

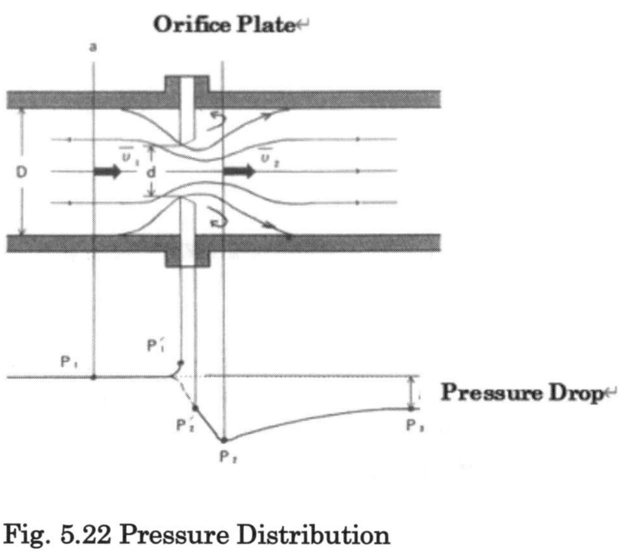

Fig. 5.22 Pressure Distribution

The differential pressure type flowmeter is still the most widely used flowmeter in the process in each factory even now. Let's briefly review the principle of the differential pressure flowmeter here.
Fig. 5.22 shows the pressure distribution diagram when the fluid flows from upstream to an orifice plate.
When the dimensions of the orifice are taken as shown in the figure, the density of fluid is, ρ_1 pressure is P_1 in the upstream cross section (a), and the downstream pressure after recovery is P_2, the mass flow rate Q_m of fluid flowing through the orifice can be represented as follows.

$$Q_m = \varepsilon C \frac{\pi/4 \cdot d^2}{\sqrt{1-\beta^4}} \sqrt{2/\rho_1 (P_1 - P_2)} \text{--------------------} (10)$$

here

β: Diameter ratio of restriction= d / D
C: Discharge coefficient
ε: Expansion correction coefficient

Equation (6) can be universally used for orifice plates, Venturi tubes, and flow nozzles, and the coefficients C and ε only change depending on the type of differential pressure flowmeter.

5.4.1 History of the Mechanism of Differential Pressure Measurement

In the case of a differential pressure type flowmeter, in addition to the mechanism generating differential pressure, the mechanism measuring the differential pressure is important, but this also has a long historical transition.

As mentioned in Section 4.1.1, it was stated that in ancient Rome at the end of the 1st century, the resistance element was inserted into an individual water pipe so that a constant flow rate would flow, but it's fun to imagine that if Roman engineers considered more, any manometer of glass tube might have been used before and after the element.

At the end of the 16th century, Torricelli became interested in the fact that Galileo said, “Pumps cannot be sucked up when the suction height is 9 m or more.” It was also discovered that the atmospheric pressure was 760 mm of mercury column in Section 2.2.2. Also, famous Robert Boyle for Boyle-Charles' law in the 17th century, fixed a long pipe for water to the outside of a four-story building built of brick, as shown in Fig.5.23.

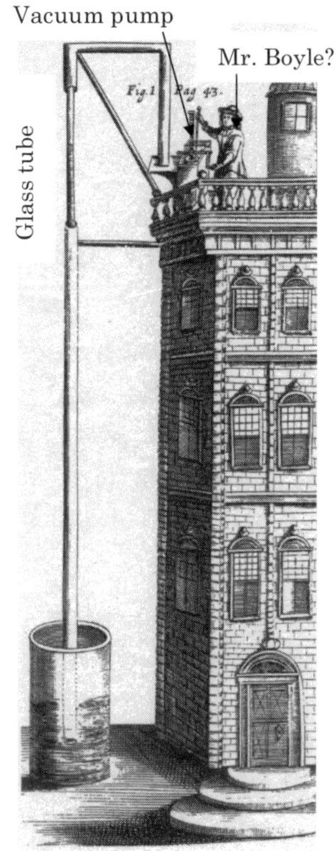

Fig.5.23 Boyle's atmospheric pressure experiment (24)

A part of the tube was made of a thin glass so that the inside could be observed. The top was connected to a vacuum pump, and when the pump was moved, the water began to rise up the pipe, but no matter how much the pump was moved, the water did not rise above about 30 feet (9.15m). The view of the experiment in a medieval building is really fun to see in this picture. Was a man at the top of building Boyle?

L.K. Spink variously classified the mechanisms of differential pressure measurements [20], but since it was not related to “When?”, such points will be supplemented with another document.

(1) Manometer [20] [26]

Herschel first introduced the Venturi meter at the end of the 19th century, but the historical fact that a water column manometer of more than 10 m was used for the experiment instead of using a mercury column manometer described in Section 4.2.4. It is probably because he wanted to improve the measurement sensitivity. (13.6 times better than mercury). In the 1910s, Waymouth used a mercury manometer with an orifice for gas flow measurement. Fig. 5.24 shows a U-tube type manometer, but it is not necessary to explain the principle. In 1932, the August issue of "Instrument" described the dial indicator type manometer for differential pressure flowmeter, and Fig. 5.25 shows one

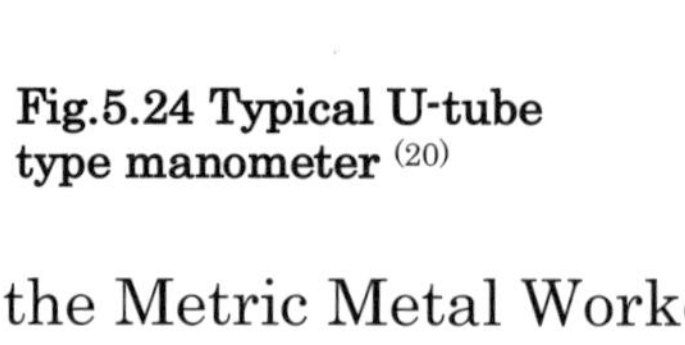

Fig.5.24 Typical U-tube type manometer [20]

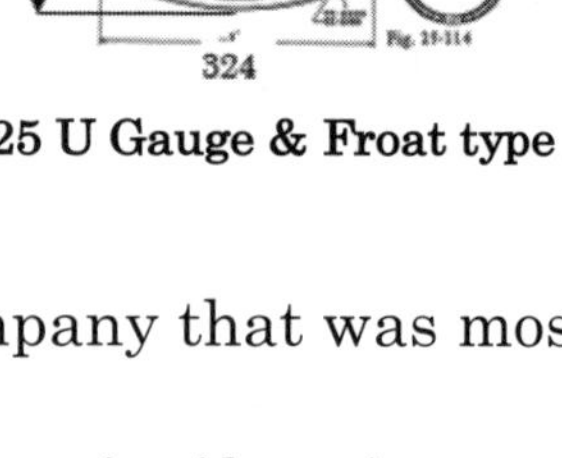

Fig. 5.25 U Gauge & Froat type meter [26]

by the Metric Metal Worker Company that was most widely used as a recorder at that time.

American Meter also made similar instruments.

It was for a maximum pressure of 800 lb (about 5.6 MPa) and prepared for the nominal differential pressures of 2.5, 50 and 100 inches. (Since the dimensions of small letters could be read from the drawing, the numerical value converted to mm are added. Pretty big indicator!)

(2) Mercury Float Type Meter [20][25]

When a lot of differential pressure type flowmeters were used, mercury float type meters had been the most used for nearly 50 years. In principle, as shown in Fig.5.26, hundreds of new flow devices appeared in the 1928~1937.

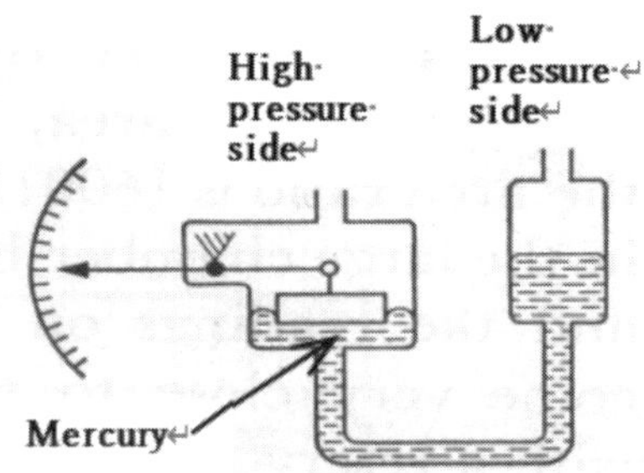

Fig.5.26 Mercury float type meters [20]

Among them, Fig. 5.27 shows the Electroflow of

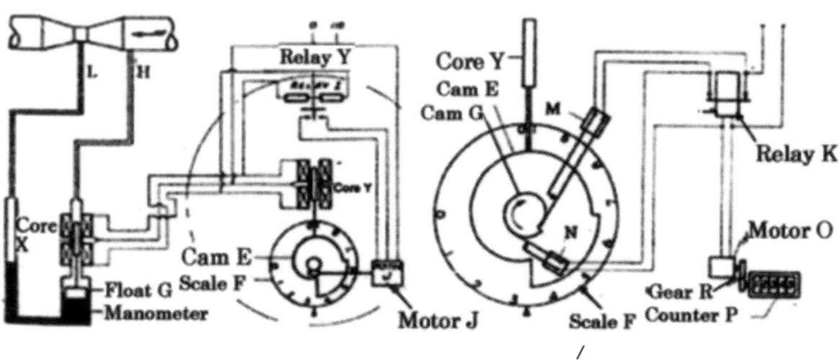

Fig. 5.27 Morey& Jones Company's Electroflow [25]

Jones Company (Morey & Jones). To explain the principle, a core X is connected to the manometer, float G of a mercury and it moves up and down depending on the magnitude of the differential pressure of the Venturi. This movement works with a core Y on the receiver side to determine the position of a square root cam E. Cam G rotates twice a minute. A contact M is fixed and starts Integration motor O. The contact N is transmitted by the cam E, also reads a scale F, to stop the integrating motor O. Relay K keeps the circuit closed after which contact M will open and until cam G will. open contact N Therefore, the operating duration of motor is in proportion to the flow rate and operate the integration counter P.

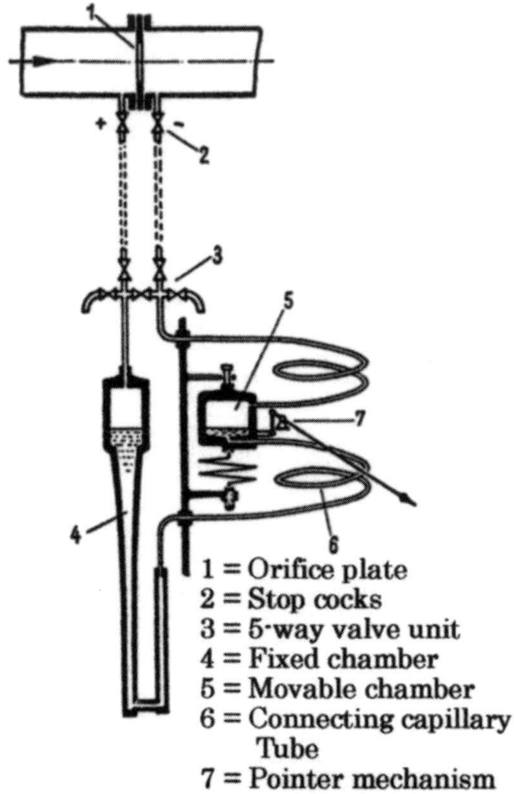

Fig.5.28 Askania secondary element [25]

Generally speaking, it can be said that this was a pioneer of mechanical-electrical integrator.

Figure 5.28 is said to be the second element (Askania secondary element) of the American Askaniya, that was introduced in the United States in 1932.

It measures the flowrate by weighing the amount of mercury that has moved from the parabolic fixed chamber (4) of the orifice to the spring-balanced movable chamber (5), since the shape of fixed chamber is parabolic, it is said that the flow rate unit is scaled equally. (7) is a pointer mechanism. However, it seems that there was a problem corrosive environment.

(3) Bell Type Meter [20][25]

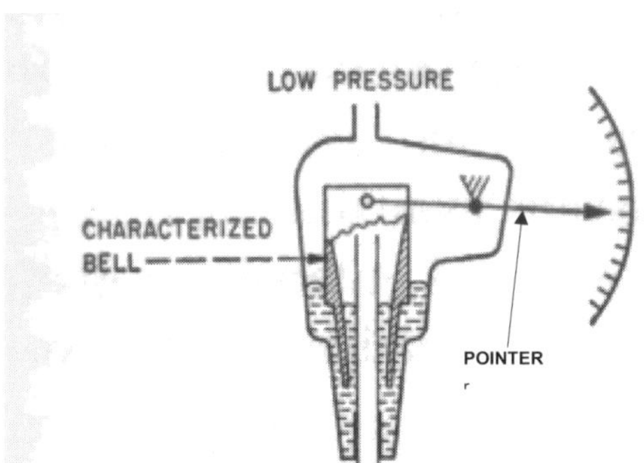

Fig.5.29 Bell type measurement mechanism [20]

A bell type meter was especially adopted when the transmission distance was long under low differential pressure, the measurement was performed at the low pressure, and a wide measurement range was required. Fig. 5.29 is it's principle diagram.

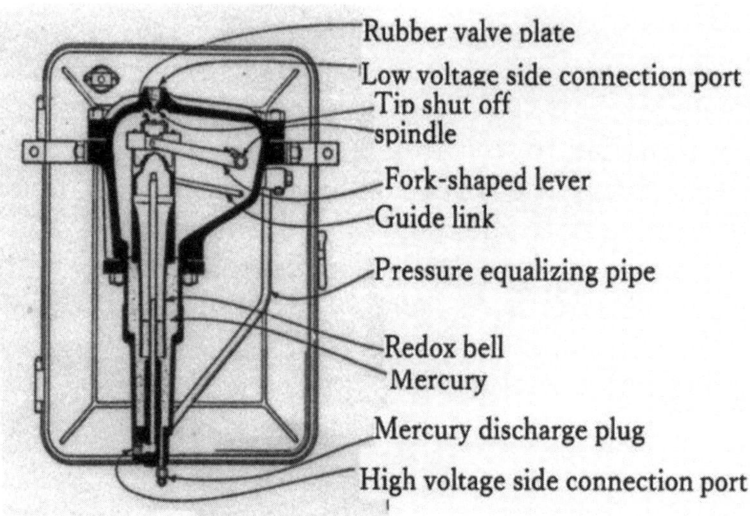

Fig.5.30 Redox bell type measurement mechanism (25)

As differential pressure increases, mercury in the bell put up side down is pushed out and mercury on the low-pressure side rises, a bell rises with the buoyancy, and a pointer swings.

The Redox bell shown in Fig. 5.30 was widely used in industrial companies such as power plants and steelworks. This type also had a square root drawn directly by a bell float, a pen arm was directly proportional to flowrate, and a chart was built into a flowrate unit. The Redox bell was used by the Simplex Valve & Meter and Belay meter.

(4) Weight Balance Type Meter [20][25][27]

A mercury weight balance type was widely used in Europe. The most popular one was the Ring Balance Meter, that principle is shown in. Fig.5.31

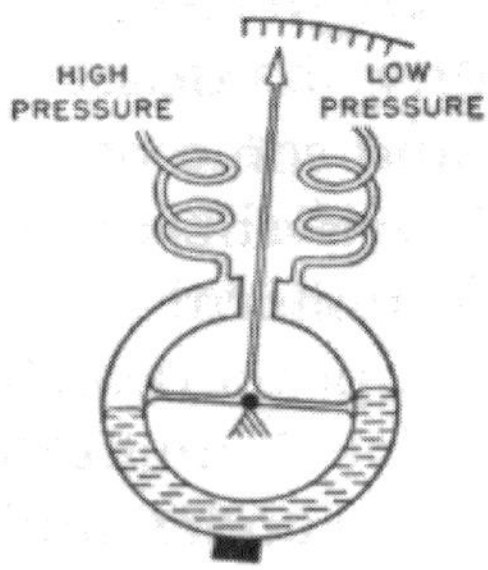

Fig.5.31 Ring Balance Meter [20]

As you can see immediately, when fluid flows, Mercury in a ring is extruded to the lower pressure side by the pressure on higher pressure side, and it rotates clockwise to maintain balance and a pointer moves. These types and ordinary mercury float meters did not have a square root output mechanism in particular, so it was necessary to connect them to an integrator that had a square root mechanism. Fig.5.32 shows a Cochrane Integrator used around 1932. To explain this mechanism, the movement of float is first transmitted to a pinion on a cam axis. The cam is molded to pull out the square root. A pin on an integrator index rises in respect to the cam so that the position of index arm matches the flowrate.

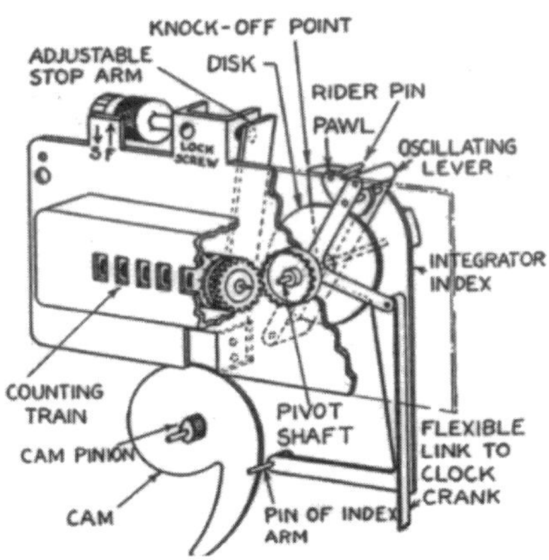

Fig.5.32 Cochrane Integrator [25]

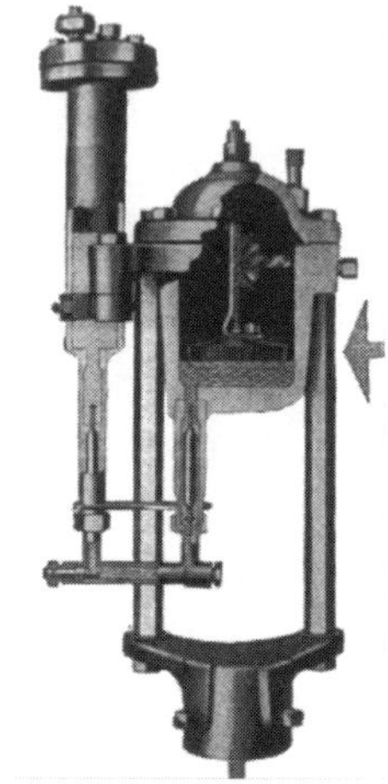

Photo 5.15
Foxborough's Mercury
Float Type meter(27

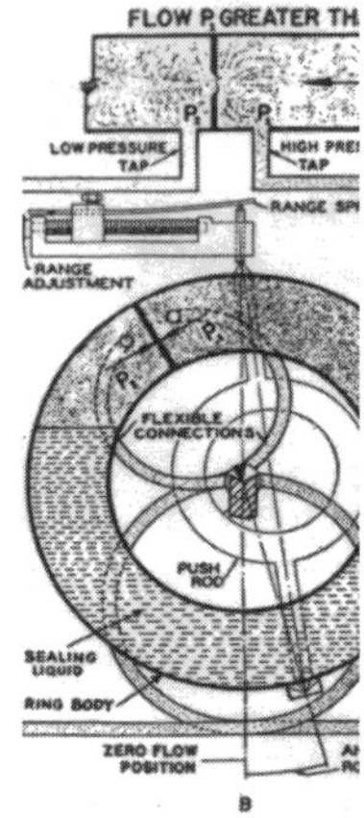

Fig.5.33 Hagan's Ring Balance Meter (27)

An oscillating lever, driven by th e clock of the instrument., carries a pawl and a pin on an integration index. When the pin reaches the knock-off point, claws engage and advance a disk mounted in the first row of counting train. On a return stroke, the claws are free on the disk. Cycles are repeated every minute.

As mentioned in a previous section, mercury was no good for the environment, but it was indispensable to measure flowrate with a differential pressure type flow meter. Even in the post-war 1950s, it was still mainstream, as Foxborough Co. advertised mercury float type meters shown in Photo 5.15 and Hagan Co. also did ring balance type meters as shown in Fig 5.33.

(5) Bellows Type Meter (20)(22)(28)

As for the mercury float type, if excessive differential

pressure was added, mercury often spouted from the instrument and was lost. Mercury consumptions were high, even if check valves and mercury pans were installed to prevent or reduce these losses.

It's not a funny story, but the mercury consumptions were not just over-ranged, it was also the work of thieves. Especially, the float-type recorders which used a large amount of mercury, so it was said that thieves stole mercury to obtain enough money as their evening job. However, this was just a nuisance, but there were various problems when using the mercury float types for chemicals. Mercury was dangerous because it formed explosive components with ordinary chemicals such as ammonia. The photographic industry was reluctant to use the mercury float method because even a small amount of liquid metal would reduce the quality of film reluctant to use mercury, the bellows type meter shown in Fig. 5.34 would be introduced. The first granted patent was a water level gauge for boilers

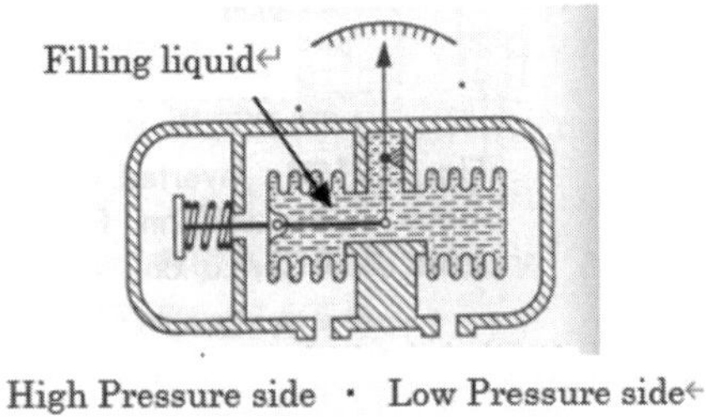

Fig. 5.34 U Bellows type meter [20]

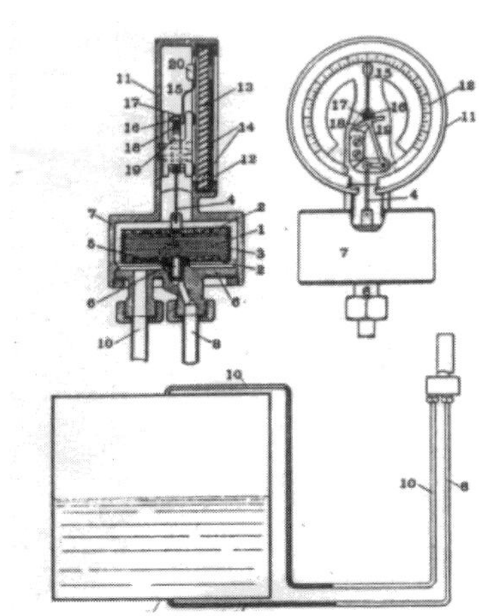

Fig. 5.35 Water level gauge for boiler by actuating Bellows [28]

operating bellows shown in Fig.5.35 in 1901. The inventor stated that the patent also could be used for flow measurement, and in fact five years later, a manufacturer was in the limelight in the water industry by marketing this type. Following patent was about a bellows type meter on a Venturi tube in 1911 as shown in Fig.5.36. It was designed to compress with increased pressure so that the bellows would not break against excessive flow rates. After that, bellows type meters were marketed by various ideas, but the influence of ambient temperature change was large and it was easy to burst, which is also the reason why Foxborough's temperature compensated mercury float types

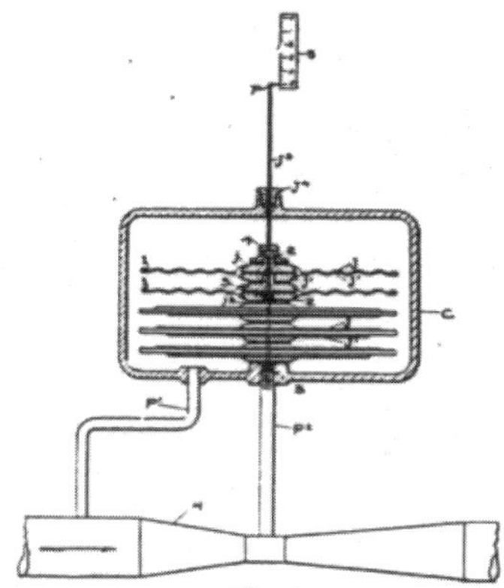

Fig.5.36 Bellows type meter used for Venturi tube [28]

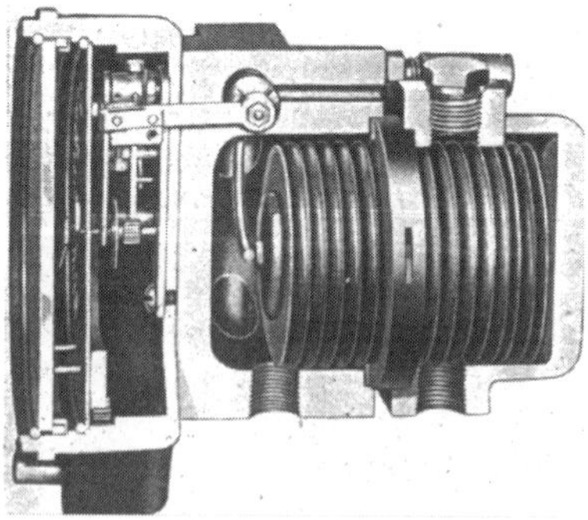

Photo 5.16 Explosion-proof bellows type meter with temperature compensation[28]

mentioned above etc. are still used. However, the first explosion-proof instrument with temperature compensation as shown in Photo 5.16 was made by Burton Instrument corp., in 1943. Looking at this, it seems that the shape was gradually becoming similar to that of modern differential Pressure transmitters.

Bellows similar to Photo 5.16 are described in detail in "Revised Flow Rate [22]" by Mr. Shozo Matsushiro as follows.

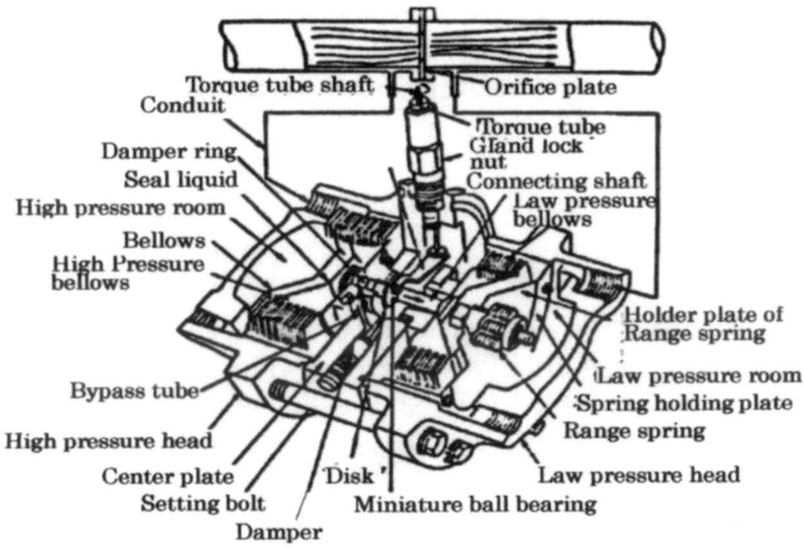

Fig.5.37 Internal structure of Photo 5.16 [22]

Fig. 5.37 is a bellows type differential pressure gauge manufactured by Hokushin Electric Works around 1958. There are two bellows in the measuring chamber, and the bottoms of both bellows are fixed to a central plate, and two bellows on both sides are connected to each other by connecting rods. Sealing liquid is sealed inside both bellows. A spring is fixed to the right end of the connecting shaft, and the other end of the spring is fixed to a central plate by a support plate. That is, the elastic forces of the two bellows are balanced, and the spring and differential pressure are balanced.

Due to the differential pressure between a high-pressure measurement chamber and a low-pressure measurement chamber on the left side, both bellows are displaced to the right, so a disc fixed to a connecting shaft pushes a drive arm via a miniature ball bearing, and a torque tube moves its rotational movement and externally transmitted. The seal liquid expands and compresses by changes in the environmental temperature, but the amount of the seal liquid escapes to the space between a free temperature compensation bellows and a high-pressure bellows via a bypass path, or is filled from there. The influence of the bellows is eliminated or the deformation of the bellows is prevented. Nowadays, the transmitters of differential pressure flowmeters have been developed to various types, so, they will be described later section.

(6) Wonderful US Oil Plants [29][30]

The differential pressure flowmeters had evolved in this way, so finally the example of orifice meters in an instrumentation in an oil plant that American Meter Co. described extensively in the 1940 "The Oil and Gas Journal [30]" was introduced.

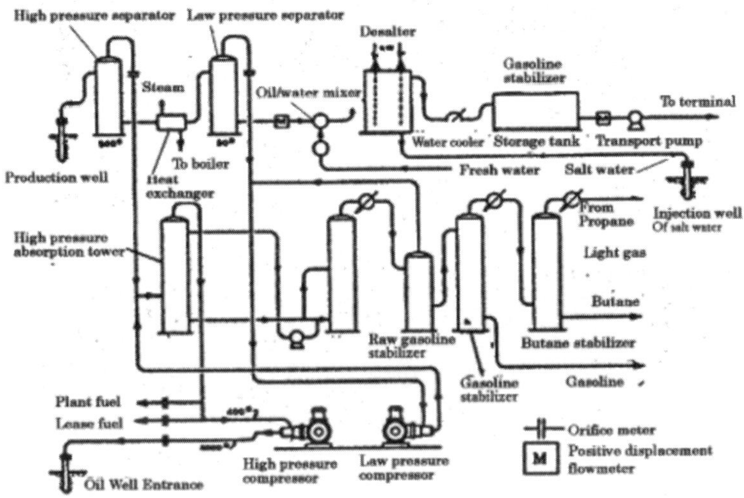

Fig.5.38 Oil and gas flow map from the Ville Platte well (29)

Fig,5.38 shows a flow chart of oils and gases to a plant from an oil well at the Ville Platte Plant (Louisiana) owner by the US Continental Oil Company.

Photo 5.17 Magnificent Group of Orifice

This high-pressure gas flows through an orifice meter to the absorption tower operating at 500 lb / in^2 (3.5 MPa). From here, it will be shipped as factory fuel or rental fuel after passing through an orifice meter.

This is what American technology was around when I was born in 1938. There were many such magnificent oil plants in USA.
Photo 5.18 shows a control room of a refinery of Imperial Oil Ltd., in Canada. Thus, automatic controls and recorders have

Photo 5.18 Control room of a refinery in the 1940s (30)

generally been accepted as media for refinery operation. In this way, the operator had to replace the hydrometer and the densitometer with a set of temperature, pressure, flowmeter readings and automatic clocks in front of him. His problem was to keep driving stable under pre-determined or optimal conditions. As much as he could, his job was to keep an eye on the instrument panel.

5.5. History of Positive Displacement Flowmeter

5.5.1 History of Oil Reclamation (31)(32)(33)

The previous section described the history of water meters and gas meters, but these meters and their improvements must have come to be used in oil trading and production processes. Let's first review how oil had been cultivated.

Photo: Woodblock print of Noah's Ark by Anthony Koberger from the German Bible

Mankind's use of oil dates back to the primitive society of using asphalt as plastic. Asphalt appears in the Old Testament in connection with the manufacturing method of Noah's Ark, but there are traces of asphalt in the farming tools found in the ruins around 7000 BC in northern Mesopotamia (now Iraq), and in ruins of Babylon City. It was also used as construction material.

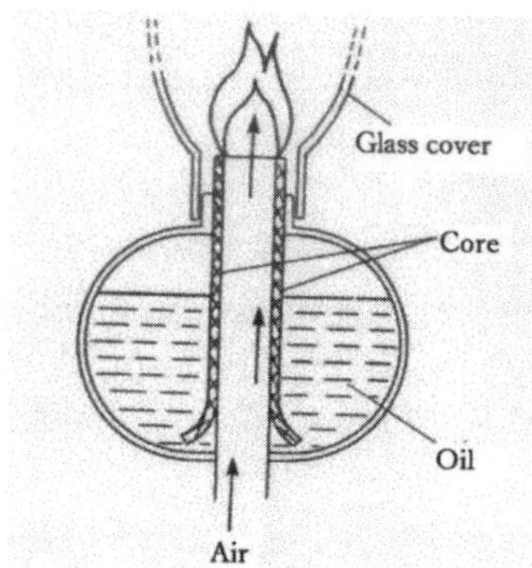

Fig.5.39 Argand lamp principle (31)

In Japan, Emperor Tenji who appeared in Section 3.2.2, in 668, when he became an emperor, Nihon Shoki mentioned about him "Koshi no Kuni (Niigata prefecture nowadays) donated burning soil and burning water to him." This is Japan's first record of oil.

The ancient lighting method that the wick in animal or vegetable oil or mixed oils is stored in plates and vases and is burned, has remained unchanged for a long time, but in 1784 Swiss Argand, a modern luminaire, invented the Argand Lamp described in Fig.5.39.
This is said to have been the greatest advancement in the history of lighting, seven times brighter than candles.

Since the beginning of the 19th century, a large amount of lighting was required, so whale oil alone which had been used until then was not sufficient in quantity. It was "coal oil" that appeared as needed.

It was James Young, an excellent scientist who succeeded in "coal oil" after helping Faraday experiment at the Royal Research Institute in the United Kingdom and later became the owner of a chemical factory in Manchester.

Photo 5.19 Drake's Well[31]

In 1847 he succeeded in extracting liquid hydrocarbons by carbonization of oily coal and oil shale, producing naphtha, kerosene, lubricants and paraffin.
Meanwhile, in the United States, E. L. Drake successfully excavated an oil well in the Oil Creek area of Pennsylvania in 1859, collecting 30 barrels (about 4,800 liters) of crude oil. Immediately, the news that "Drake's folly" was successful spread and became famous as "Drake's Well". The person in a silk hat in Photo 5.19 is E.L. Drake. The success of Drake's well caused an oil boom and the oil industry was born.

Photo 5.20 Amase Oil Field of Japan Oil Company (31)

It was John D. Rockefeller who established the oil industry as a modern business. He founded the US Ohio Standard Oil CO., in 1870, and in 1878 the amount of oil refining capital investment accounted for 90% of the total United States.

The domination of Standard Oil in the United States also meant a monopoly on the world oil market. Until the advent of Russian oil in the 1880s, American kerosene, which dominated the world and was under control of Standard Oil in Japan, controlled almost all of the oil market.

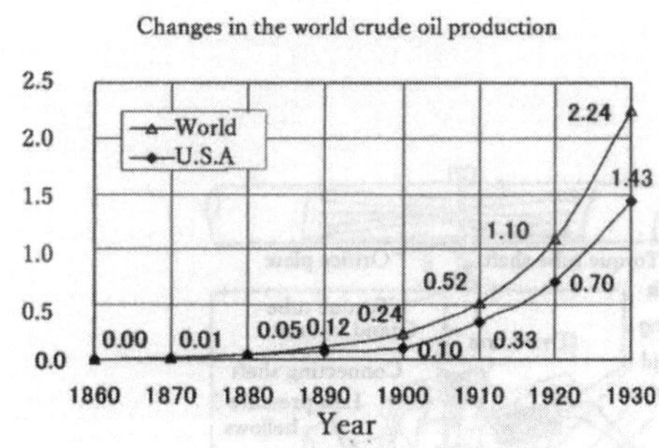

Fig.5.40 Change of world crude oil production output (33)

Around 1885, Hisahiro Naito in Niigata saw people flocking to Amass for oil well drilling and became interested in the oil business, and established Japan Oil Company on May 10, 1888.

Photo 5.20 is a photo of the Amase Oil Field at its peak in the middle of the Meiji period when a lot of oil wells and Yagura (turrets) stood there.

As shown in Fig.5.40, the production output of world crude oil had increased rapidly in the 1900s.At present, the annual world crude oil production output is about 4 billion kl. It was about 1/18 of that in the 1930s. By this time, American oil production output accounted for 64% of the world output.

5.5.2 Positive Displacement Type Oil Meter (34)

The oil industry became active in the United States, and as mentioned in section 5.4, many oil plants had been built. As shown in the flow chart, orifices were used for the gas flow measurement of these plants, also positive displacement meters were used for the oil flow measurement.

The title was meant to be "complete discharge type oil meter", but about this definition and the types of volumetric flowmeters. I would like introduced here, as it was featured in "Instrument" from the November issue of 1932 to the, December issue of 1934.

In 1916, the ASME Flowmeter Expert Committee made a plan for research on positive displacement flowmeters. Around 1920, "positive", which had been used indefinitely until then, was defined as "Fluid cannot pass through a meter without moving the main equipment." In the industry these days, "monkey wrench and screwdriver mechanics" was a way to give technicians the ability to study to be technicians on their own, with only vague classifications and sloppy terms from the old days. However, these days, the definition as “a positive meter is a quantity meter, that is, a quantity counter.” had been taken root. Table 5.1 shows the classification of positive displacement flowmeters at that time.

- Positive Displacement Flowmeter (Type)
  - Reciprocating Piston (Type)
    - With crank
    - No crank
    - Nutating Disk
  - Oscillating Piston
  - Rotary Piston
  - Nutating Disk
- Gas Meter
  - Dry Type
    - Rotary
    - Bellows
  - Wet Type
    - Drum
    - Nutating Bell

Table 5.1 Classification of positive displacement flowmeters at that time (34)

(1) Reciprocating Piston Type Meter

Photo 5.21 shows the internal s tructure of the Baslller meter of the American Liquid Meter Co. This was a new form of the old crankless single piston, specifically designed for petroleum products.

Photo 5.21 Internal structure of Basller

Fig.5.41 shows the operating principle that the piston reciprocates toward one of two main valves at both ends.

reciprocates from side to As the piston side, a spiral shaft activates a pilot valve as well as the counter of an integrator on the upper left side. A spiral shaft rotates an end gear to move the pilot valve left and right to switch between two passage ports (Inlet and Outlet) at different positions above and below the inner cylinder so that fluid is discharged from an inlet to an outlet on the way back and forth from the piston. (Even if you read the above explanation, you will not fully understand the mechanism because it is very difficult. If necessary, please check out reference No.35)

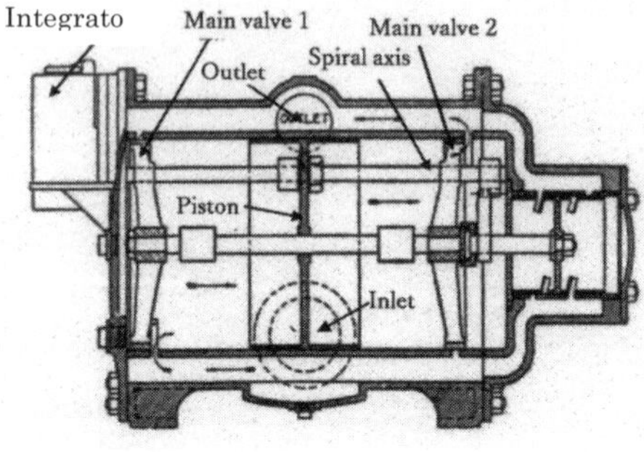

Fig.5.41 Operating principle of Basller meter (35)

The bore size is 1 to 6 inches and the capacities range from 2 to 25 GPM to 125 to 1250 GPM (0.45 to 5.7 m^3 / h to 28.4 to 284 m^3 / h), respectively. Fig.5.42 shows the Pittsburgh oil and gasoline meters

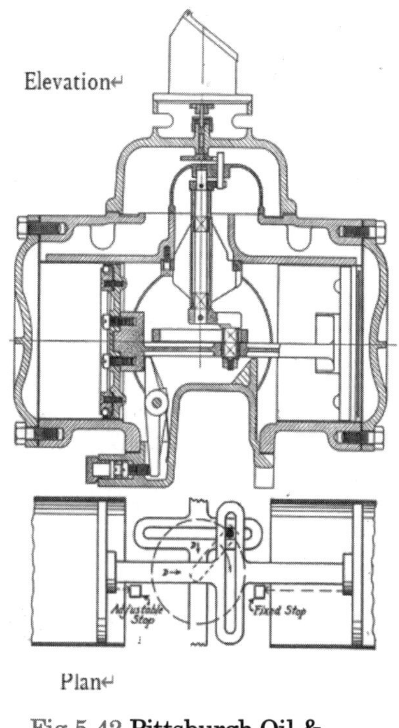

Fig.5.42 **Pittsburgh Oil & Gasoline Meter** (35)

manufactured by Pittsburgh Equitable Meter Co. at the same time. This meter carries a rotary valve and uses a type of Scottish yoke to drive an integrator.

This meter is also a piston meter, and the mechanism shown in the plan view is the Scottish York, and when the piston moves left and right, the clockwise rotation is taken out and transmitted to the integrator. With a size of 2 inches and a maximum capacity of 120 GPM (27.3 m^3 / h), the error can be measured within 0.4% even at 1 GPM, and within 1% over the entire flowrate range, which was a good performance at that time.

Fig.5.43 shows Badger G.P. of Badger Meter Mfg. Co.'s It is described in detail in reference [35], but as you can see, the internal unit has a very complicated Mechanism and may be difficult to understand. I'll try describing here. As it can be seen from the figure, there are 4 pistons and 4 rollers and many other parts, therefore, maintenance will be difficult. The appearance displacement flow meter must be awaited.

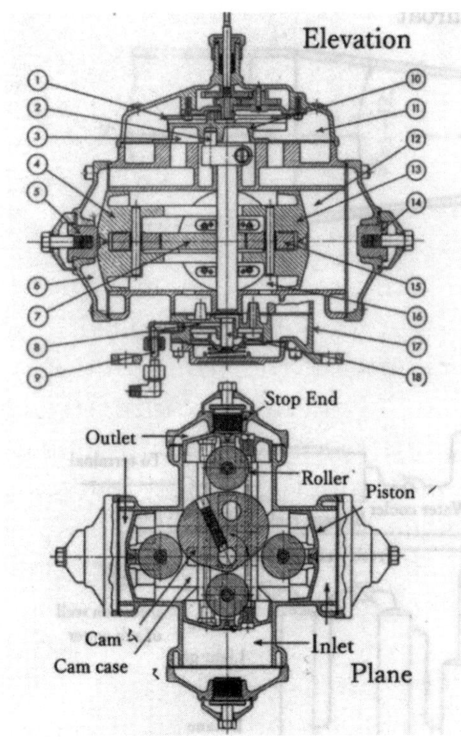

Fig.5.43 Badger Meter [35]

This meter was usually manufactured in 1/2 inch (8GPM = 1.8 m^3 / h) and 2 inch (125GPM = 28.4 m^3 / h). (Since there was still little leakage, it was suitable for measuring small flow rates and it is said that many meters were used in gasoline measuring instruments.)

Photo 5.22
Basller meters (above) and Pittsburgh Meters (Bottom) that were used in large quantities (29)

Any way, it seems that Basler meters and Pittsburgh meters were often used to measure liquids in American oil plants at that time, and Photo 5.22 shows a group of meters used in large quantities in the Ville Platte oil plant introduced in the previous section.

Oil from the low-pressure separator was being measured and the meters were compensated by temperature and converted to the volume of 60 ° F. The meters were installed for constant checking and were equipped with meter calibration devices consisting of tanks that were accurately calibrated for precise calibration at any time and pumps that could circulate the actual amount to meters.

Master meters were also already in use for checking.

(2) Rotary Piston Type Meter

Fig. 5.44 shows the operating principle diagram of the Rotary Piston type meter.

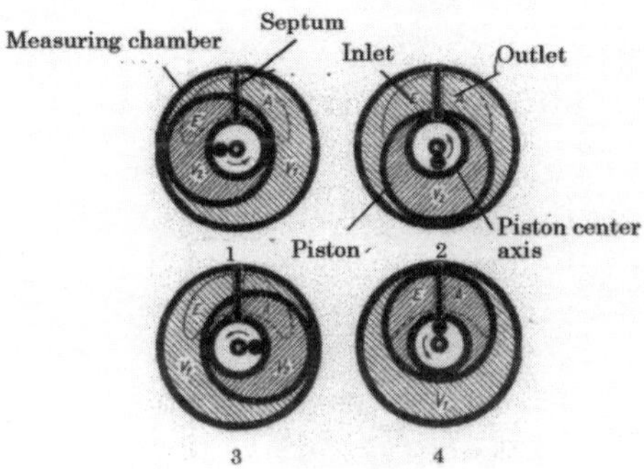

Fig.5.44 Operation principle diagram of rotary piston type

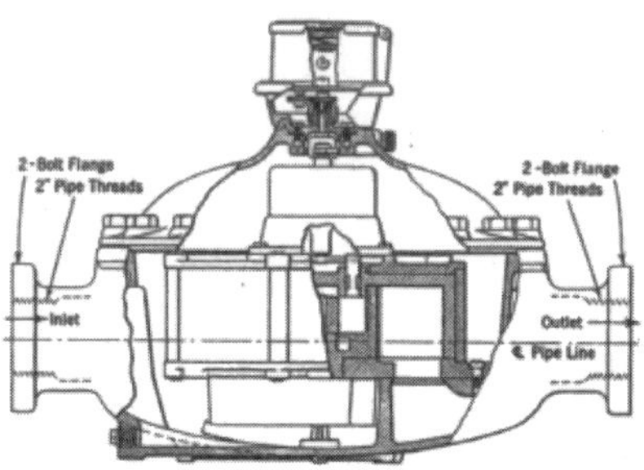

Fig.5.45 "Empire" Meter of National Meter Co.(37)

There is a septum in the radial direction of a measuring chamber, and a piston is eccentric with respect to the center of the measuring chamber. A notch provided in the piston sandwiches the Septum. When the fluid flows in from the inlet, the piston rotates in the order of 1 → 2 → 3 → 4 by fluid pressure and pushes fluid out from an outlet, but the notch reciprocates on the septum. This is the mechanism of the rotary piston type meter.

Fig. 5.45 shows a rotary piston type meter released by National Meter Co. as a water meter under the name of "Empire" in 1884. It had been developed for many liquids such as oil, ammonia, beverages, ink, syrup, tar and varnish.

(3) Nutating Disk Type Meter

The word “nutate” is synonymous with sway, rock, nod etc., but in technical terms, the spinning top movement is called as "precession movement", which means the nutate. It is said that the axis of rotation of the earth also makes one cycle of nutation in each 25,000 years.

The nutating disk type meter should be imagined as the swinging motion of spinning top.

Photo 5.23 shows the 3D view of the nutation disc type meter.

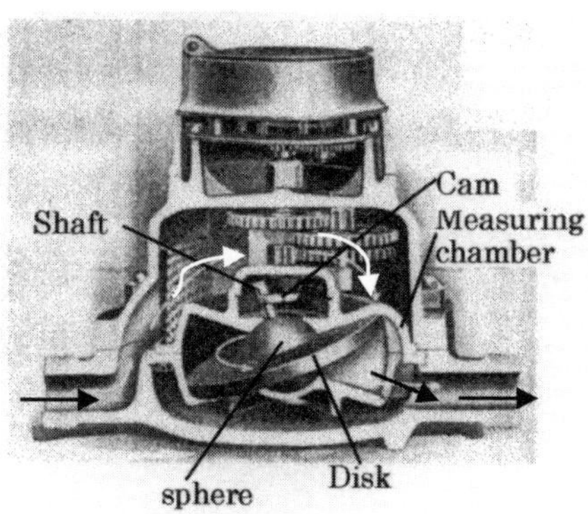

Photo 5.23 Nutating disk type meter (36)

The upper and lower surfaces of measuring chamber are conical toward the inside, and the side wall of the inner chamber are spherical. A disk that is a movable element is embedded in the central sphere, and a pillar or axis perpendicular to the disk extends out of measuring chamber. The inclination of the axis is kept by a cam so that the disk is in contact with the bottom of the measuring chamber along a radial element on the side of sphere.

An inlet and an outlet are on the side wall of case, opposite to the radial part. The liquid enters the measuring chamber alternately from the top and bottom of the disk, and passes around the conical measuring chamber to the outlet. Since the liquid first applies pressure to the upper side of the disk and then to the lower side around the measuring chamber, the disk causes nutation. The disk has a groove in the radial direction so that it meshes with a partition of the measuring chamber to prevent rotational movement. Therefore, the disk only makes nutation, that causes the top of shaft to make a circular motion and engages with a crank on a pointer shaft to move the pointer of this meter.

All of the positive displacement flowmeters mentioned so far had been complicated piston types or its variations. However, from here, it shifts to the rotor type flowmeters which may be easily understood.

(4) Roots Gas Meter

As for gas meters, the drum types have already been mentioned in Section 4.1.1(4), but finally the roots gas meter appeared. The appearance of a two-leaves rotary pump introduced in 1902, was patented in 1924 under the name of Roots Meter. Initially used only to measure low pressure and clean lighting gas, various models were made in a wide range for industrial use in the 1933. Fig. 5.46 shows the principle of the roots meter, but the rotors are not in contact with each other, there is a slight gap, and there are two pairs of timing gears (indicated by a dashed line) rotated at both ends on the outside of the

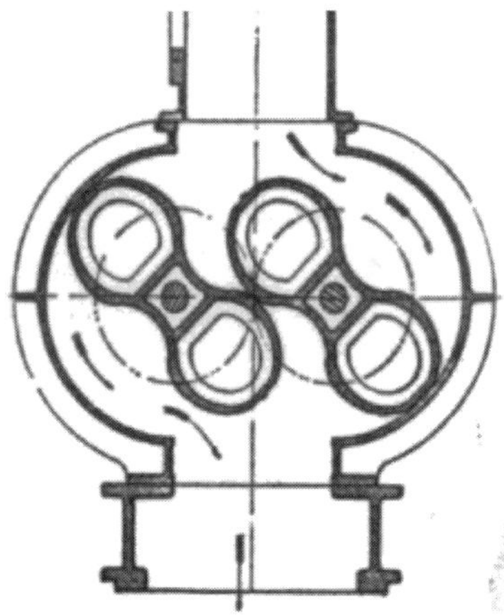

Fig.5.46 Principle diagram of Roots meter (38)

Photo 5.24 is a Roots-Connersville-Wilbraham equipped measuring chamber.
with a PVTT recording gauge that measures gas with large

Photo 5.24 Cornusville Roots Gas Meter (38)

changes in pressure and temperature and records the transition of pressure, volume, and temperature. This chart was turned by oneself instead of a clock, and it was possible. to record the average value once a day.

(5) Oval Meter (39)

By the way, the oval gear type flowmeter appeared in Japan around 1950 after World War II, but when was it first available in the world?

According to Bopp & Reuther of Germany, the first oval meter was made in 1933. Fig. 5.47 is a principle diagram of an oval gear type meter, which has a large number of teeth cut into an ellipse, and the left and right rotors were engaged with each other to rotate. Therefore, unlike the roots meter, timing gear is not necessary. The shaded area surrounded between measuring chamber and oval gear in the figure becomes a measure, and one rotation discharges four times this amount. Photo 5.25 is a photograph of the measuring room and two oval rotors taken from above a cover plate. (Oval Corp. offered this after World War II)

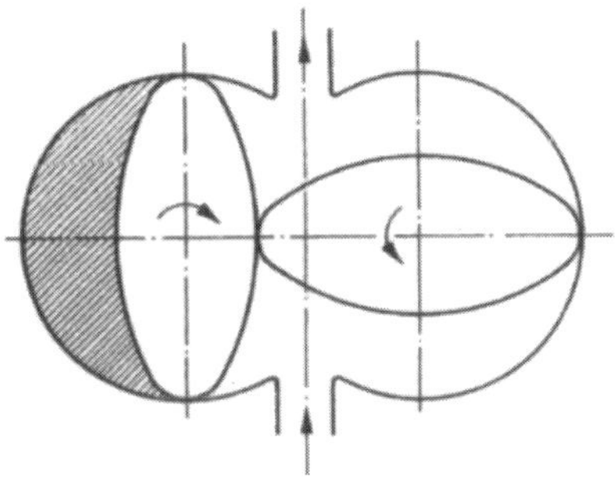

Fig.5.47 Principle diagram of oval meter (39)

Photo 5.25 Measuring case of Oval meter (39)

(6) ASME Flowmeter Expert Committee Established

There is no evidence that oval meters were used in this era in the United States. About seven years later, in 1940, ASME's Flowmeter Expert Committee decided to focus more or less on the meters used in the petroleum industry. (37)

The first reason was that the number of meters used in the petroleum industry had been steadily increasing in the last few years, and the second reason was that the use of meters in several gathering lines and piping lines were Required.

These were due to the facts that the legislative departments of some states had been discussing the law in various ways.

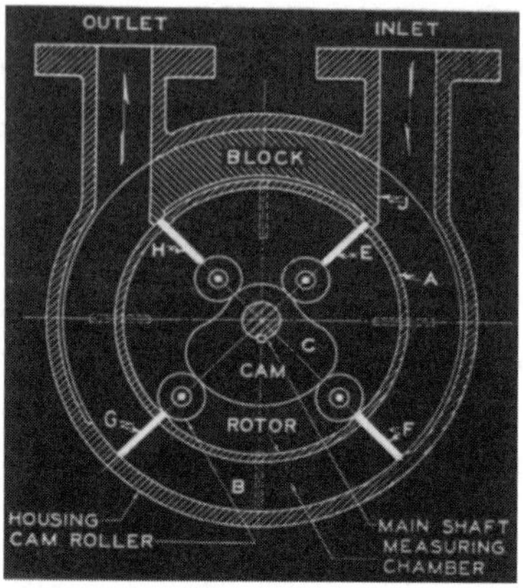

Fig.5.48 Sliding vane type meter (40)

As a result, the Commission was convinced that it should provide fair information to meter manufacturers, meter users and general citizen through a competent authority responsible for measurement of retailing and wholesaling petroleum products.

Therefore, they prepared equipment to test meters used in oil wells, transportation lines and refineries. Next, it was necessary to investigate the meters used in the retail sale of products such as fuel oil, gasoline and liquefied petroleum

The types of oil meters were almost the same as seven years prior, but only sliding vane type (see Fig.5.48) were newcomers in the USA.

5.5.3 Calibration of Positive Displacement Flowmeters for Measuring Volatile Oil (40)(41)(42)(43)

The ASME Flowmeter Expert Committee mentioned in the previous section, announced the calibration results of the flowmeters for measuring volatile oil such as gasoline as early as June 1940.

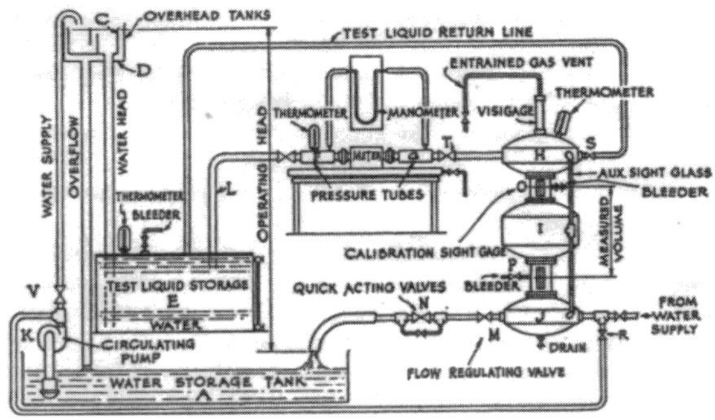

Fig.5.49 Schematic of equipment used for meter testing (41)

In order to calibrate these flowmeters, consideration must be given to reducing errors due to evaporation of volatile oil, temperature changes and gas contamination. Therefore, the test was conducted using the calibration device shown in Fig. 5.49 which was patented by the Gulf Research & Development Company.

This test device was very elaborate so that any air contamination was eliminated. Water in the Water Storage Tank A is pumped up to an Overhead Tank C by a circulation pump K and then dropped into Test Liquid Storage Tank E in which the water is stored in the lower part of the tank E and test liquid is stored in the upper part, the test liquid is pushed out through a pipe line L. It passes through a meter under test and enters an upper tank H of a prober. The water and the test liquid are separated at the position of the upper glass gauge O of the prober, and water is contained below it, but the water passes through a flow control valve M and a rapid opening valve N and returns to the Water Storage

Tank A. If the rapid opening valve N is closed when the boundary layer between water and liquid reaches the lower glass gauge P, the volume of a measuring liquid is from O to P, and an error is calculated by comparing it with the indicated value of the meter under test.

Of course, a thermometer is attached near the prober and the meter to correct the temperature expansion of the liquid. The pressure loss of the meter is also measured upstream and downstream of the meter with a manometer as you can see. The accuracy of this test device is estimated to be 0.1%.

Now, two types of piston type flowmeters were tested using this prober. One was a pack shield meter and the other was a film shield meter.

The pack shield meter is one in which a leather, synthetic or metal ring is embedded in the piston and is tight to its casing, while the film shield meter has a slight gap between the piston and the casing. There is the thin film of liquid in the gap. There is no description of these meter sizes, but it will be about 2 inches in terms of flowrate.

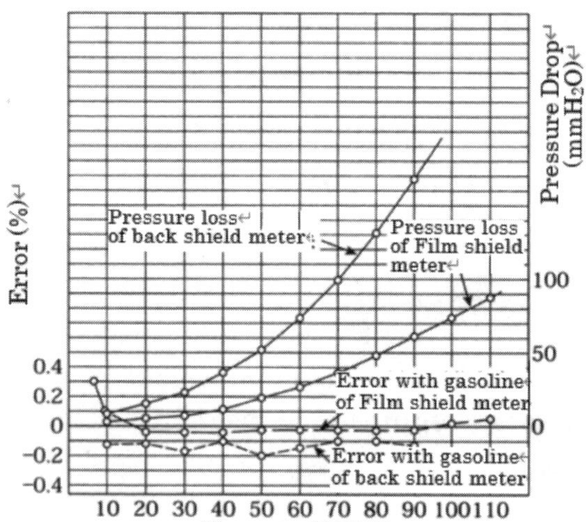

Fig.5.50 Basic characteristics of piston meter (40)

To introduce a few test results, Fig. 5.50 shows the basic characteristics of both meters of gasoline. The definition of error (%) is the opposite of the current definition of instrumental error because it is the value (%) obtained by subtracting the reading amount of tested meter (I) from the passing amount (Q) between O and P of the prover dividing by the passing amount.

That is,

$$E = (Q - I)/Q \text{ -- (11)}$$

But nowadays, the meter error is defined as follows

$$E = (I - Q)/Q \text{ -- (12)}$$

If Error (E) is calculated by equation (11), the more positive it is, the larger the leakage of the meter. Looking at this figure, since the piston of the pack shield meter is in close contact with the casing, the pressure loss is naturally larger than that of the film shield meter. However, the accuracy with gasoline has very good line arity within 0.2%.

The next Fig.5.51 shows the accuracy result after durability test of the film shield meter.

In this figure, ① is the result of calibration with the prober of manufacturer, and ② is the result of

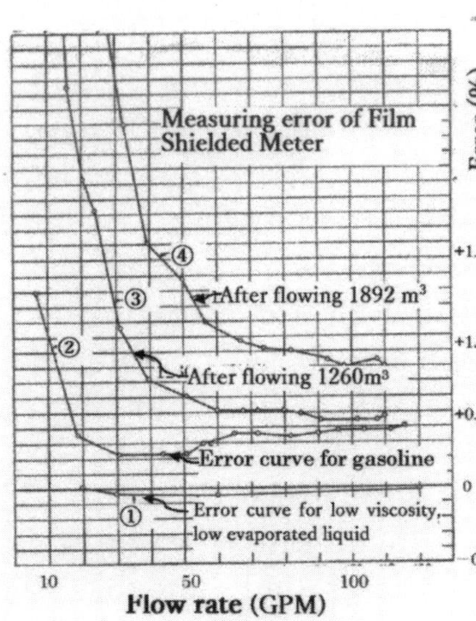

Fig.5.51 Accuracy result after durability test(40)

calibration with gasoline using the test equipment shown in Fig. 5.49.
There is a deviation of about 0.4% between ① and ②. To make matters worse, as the amount of gasoline flowing through the liquid increases, the error between ③ and ④ increases, the instrumental error at small flow rates became particularly worse.
The author attributed this to the wear of the meter elements and integrators, and the increase of loaded torque. Fig.5.52 shows the viscosity characteristics of the back shield meter.
The viscosity of gasoline and furnace oil is different by about 2 cP, and the instrumental error shift is about 0.5 to 0.6%, but it seems that it is not so bad compared to current positive displacement types. It's just amazing that there was a system in the United States to calibrate such marketed petroleum flowmeters, 80 years ago during World War II. In this way, positive displacement flow meters (PD meter) had come to be widely used for measuring volatile oil, and in particular, reciprocating piston meters such as the

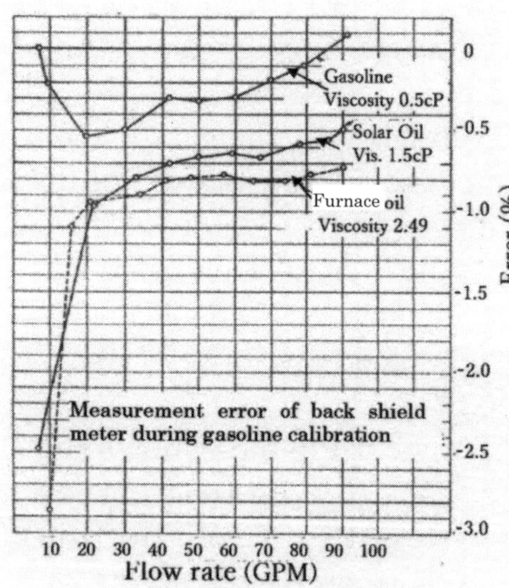

Fig.5.52 Viscosity characteristics of piston meter(40)

Pittsburgh meter were often used for gas stations, and after World War II, about 100,000 units had been almost exclusively delivered every year.
In the 1950s, "Rotocycle" of Rockwell Co. and "Brodie Meter" of Brodie Co. entered the market in the United States. For large-capacity meters such film- shielded type with low pressure loss occupied the mainstream.

Photo 5.26 is a Rockwell's tank truck meter, and Fig.5.53 is its principle diagram. As shown in the figure, four semi-cylindrical blades are embedded in the central rotor and rotate freely with respect to each axis, but all four mechanically maintain the same orientation.

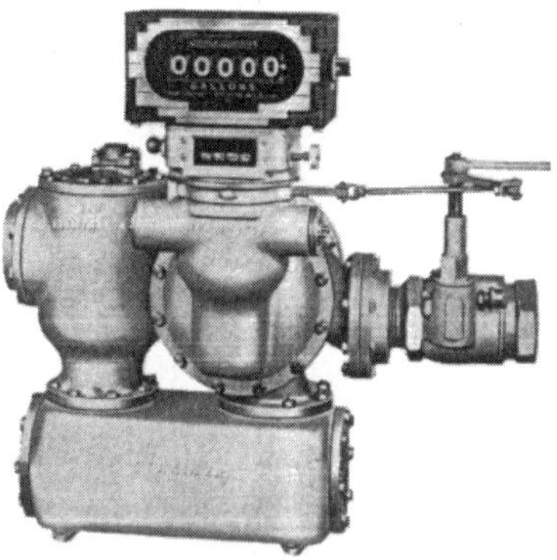

Photo 5.26 Flow meter for tank truck of Rockwell Co., (42)

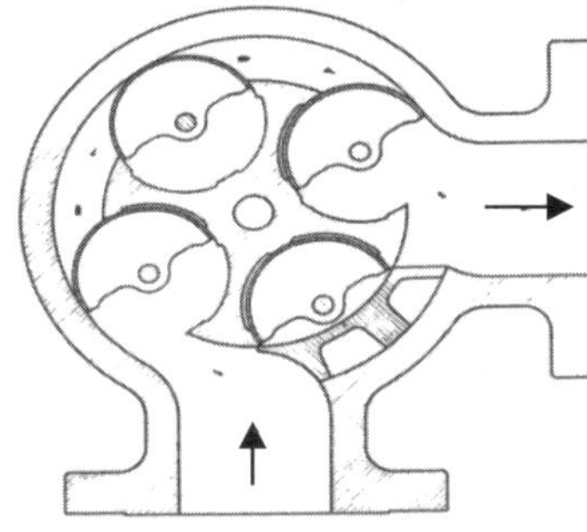

Fig.5.53 Principle diagram of rotocycle meter (42)

Photo 5.27 Brodie Meter (43)

At least two blades always work as seals. Photo 5.27 is a Brodie meter, and Fig. 5.54 shows two rotors engaged inside.
In this way, PD meters had become mainstream from reciprocating type to rotary type.
Furthermore, the calibration method was also standardized close to recent ones.

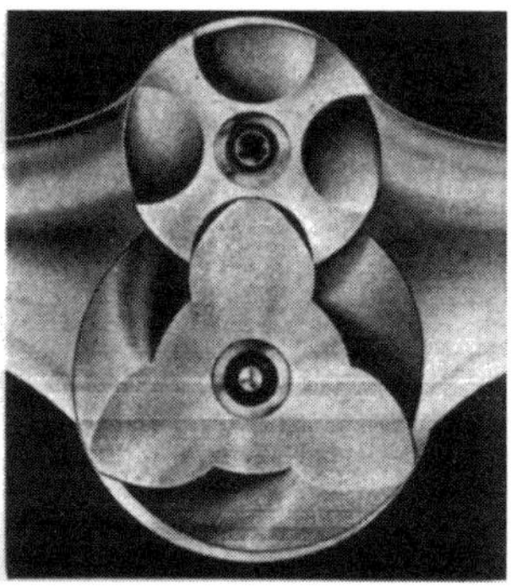

Fig. 5.54 Rotors of Brodie Meter (44)

5.5.4. History of Gas Stations

Automobile is derived from the fact that the Daimler and Benz of Germany separately made gasoline engine cars in 1886, about 135 years ago. Then in 1903, Ford started the "Ford Motor Company" in Detroit in the United States. Photo 5.28 is Daimler in his later years. Photo 5.29 shows the Ford Model A that was Ford's first automobile.

Photo 5,28 Daimler in his later years

After that, with the entry of General Motors and Chrysler, automobiles began to be mass-produced, and the number of automobiles produced in the United States increased dramatically.

As shown in Fig. 5.55, the number of automobiles used in the United States shows rising growth, reaching 10 million in 1920 and 32 million in the 1940s.
Therefore, as mentioned earlier, it is no wonder that the number of PD meters for refueling these vehicles had

Photo 5.29 Ford Model A that is first car (44)

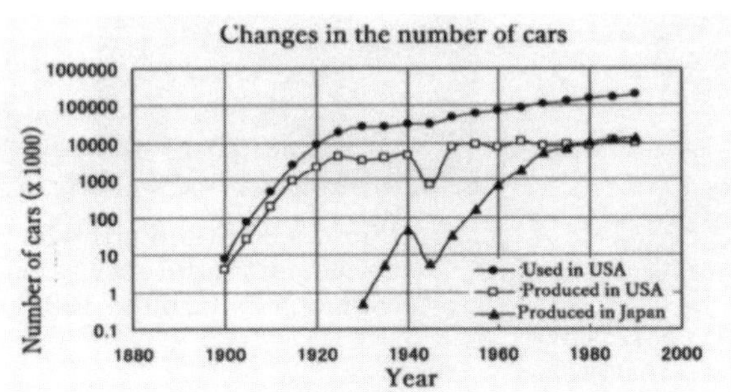

Fig.5.55 Changes in the number of cars (45)(46)

increased by 100,000 every year.
As shown in Fig.5.56, female drivers already appeared in Japan in the early Showa period (1920). This is a commercial picture of a MOGA (modern girl) driving a car by Nippon Oil. Photo 5.30 is a scene where Japanese stand girls were Co. playing an active role in visible gasoline stands in the intense sales competition around 1933.
Any literature about gas stations in the United States was searched for, but nothing could be found. However, as a valuable document in Japan was found, I would like it to be. introduced first

Fig. 5.56 Appearance of female driver (early Showa period) (31)

Photo 5.30 Stand girl who played an active part in intense sales competition (around 1933) (31)

According to the "Sekyu Jihou (Oil Times)" issued in July 1930, a pump was first installed in Japan for the purpose of selling gasoline in around the Taisho period 6 (1917) that Former Sale Fraser Company imported from the United States.

It was as simple as sending the gasoline in an underground tank to the tank of

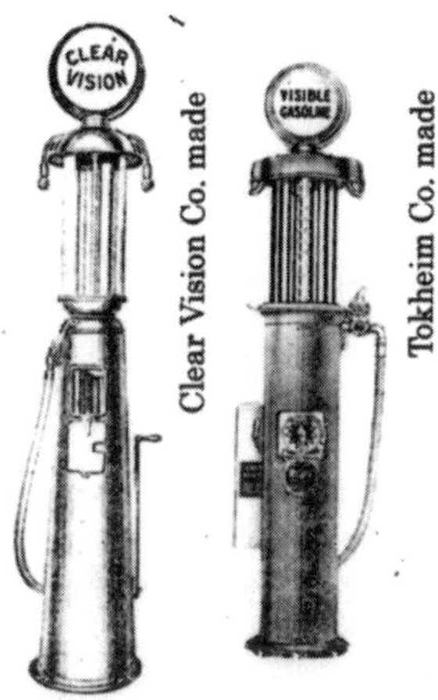

Photo.5.31 Visible stand (47)

Supplier of fixed stands in Japan

There are three types, made in Japan, made in the United States, and made in the United Kingdom, the outline is as follows.

Japanese products
- Tominaga MFG Co.:products (Conventionally mainly in Kansai)
- Tatsuno Corp.: products (Conventionally mainly in Kantou)
- Tyousyu Syoukai: product (Mainly for private use around Tokyo)
- Mituyama Kikai Seisakusyo products
  Uraga Dock Co.: products
  Other's :product
  (Recently, going to this area)

Imported products
- Milwaukee Co.product (Nihon Jidousya: Agent sales)
- Tokheim Co. product (For U.S.A product:Mitubishi Syouji Co. handling)
- U.S.A sales Co.:products (Rising Sun Oil Co. as private cars)
- U.S.A, G.B Co.: products (Standard Oil Co. as private cars)
- Pauza Co. products (U.S.A)
- Deton Co. products (U.S.A)
- Shell Mex Co. products (England)
- Other's

Fig.5.57 Supplier of fixed stand in Japan (47)

automobiles.

Around February 1919, at the suggestion of Tokyo Gas Electric Engineering Company, the Japan Oil Company installed a visible stand at Okuda Shoten on the Kamakura coast in Kanda-ku, Tokyo (currently Sotokanda, Chiyoda-ku). There was also a photo of the "Evolution of gas stations in the United States" at that time, but unfortunately the age was not added, but Photo 5.31 shows the visible stands made by Tokheim Oil Tank Pump Co. and Clear Vision, probably from the 1917s. In addition, there was an article on "Suppliers of Japanese fixed stands". Since this is written in Japanese, the translated English one is shown as Fig. 5.57. At that time, in addition to imported products, domestic products were already handled by the Tatsuno Corporation and Tominaga Manufacturing Co., Ltd. So, I would like to touch upon the histories of these companies.

5.5.5 History of Japanese Gasoline Measuring Instruments

Uchu Tatsuno was born in Imogawa, Kamiminouchi District, Nagano Prefecture. He had been very talented since he was in an elementary school, and he taught students on behalf of the teacher, when there was no teacher.

Photo 5.32 Profile of Ucyu Tatuno (50)

He didn't want to be in agriculture and tried to go to an industrial high school, but his parents didn't allow him and he went to Tokyo for an apprenticeship. After a while, after becoming a craftsman at an ironworks, he had a hard time studying at night. However, in 1911, he decided to do something and bought a storehouse in a corner of a restaurant in Matsumoto-cho, Shiba-ku, Tokyo for 150 yen in cash. This was the beginning of Tatsuno Seisakusho, which established a manufacturing plant for gas meter parts and accessories.

In 1914, he started to trial the production of gasoline measuring instruments and underground safety storage devices. Japan's first visible gasoline measuring instrument

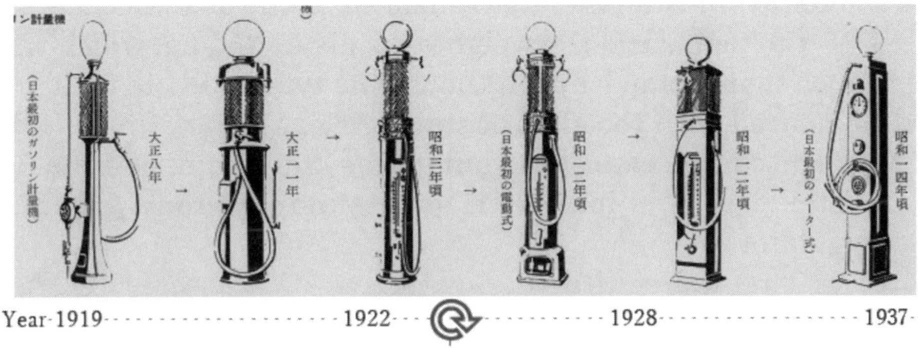

Fig.5.58. History of Tatuno-made Fixed Gasoline measuring instruments (48)

was completed in 1918, and the one of Tatuno-made meter type was first commercialized around 1939. Fig. 5.58 shows the development process of the fixed gasoline measuring instrument published in the history of Tatsuno. The gasoline measuring instrument surrounded by two women in Photo 5.30 seems to be the third one from the left in Fig. 5.58.

An inside drawing of first meter type in Japan is shown in Fig. 5.59. Tatsuno Mechatronics Co., Ltd. provided us with valuable drawings. Thanks so much about this.

The pump ④ is rotated by the electric motor ② to suck gasoline from the suction pipe ①, and dispense it into

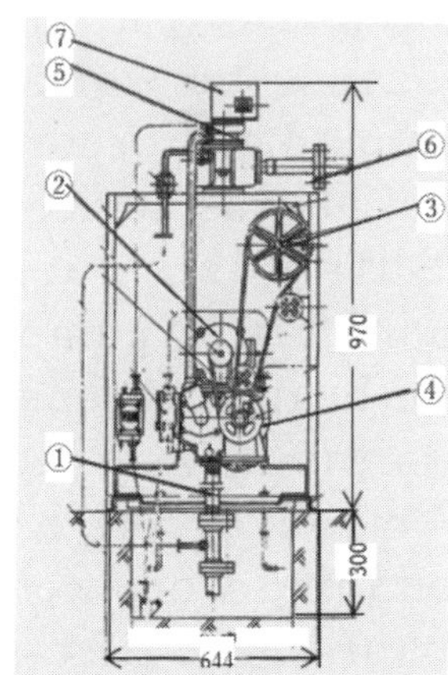

Fig.5.59 Inside of meter type gasoline measuring instrument

The tank of the car from the discharge port ⑥ via the flow meter ⑤. ⑦ is a flowrate transmitter. The size was 644 x 287 x 970 mm and the weight was about 150 kg, which was much more compact than the visible type in Photo 5.30, if compared with the sizes of stand girls.

The discharge rate of the pump was 40 L / min, and the flow meter was equipped with a Tatsuno's piston-type micro regulator.

On the other hand, Tominaga Seisakusho is a company founded by Tadajiro Tominaga on May 5, 1887. His father, Tadashi, was close to Shimadzu Genzo, the founder of Shimadzu Corporation, partly because he worked for Shimadzu Corporation, and he was advised that "Shimadzu does not manufacture pumps and heating, so why don't you do them?" For health reasons, he introduced Tadajiro early to work in his business. Tadajiro decided to start a business on Tominokoji-dori, Kyoto. The main products at the time of its founding were wood-burning stoves, pumps, water supply fixtures, etc. It is said that the sales of wood-burning stoves were particularly popular in Kyoto where it was biting cold.

Photo 5.33 Tominaga-made Gasoline measuring instrument in 1915[49]

It was in 1915 that he began to manufacture the gasoline measuring instrument. In 1918, he succeeded in manufacturing "a glass cylinder type gasoline portable weighing machine for 5-gallon, 20L" with reference to the gasoline measuring instrument imported by the importer Rising Sun (See Fig.5.57). Photo 5.33 is a visible gasoline stand manufactured by Tominaga Seisakusho around 1915, and the second person from the right is Shotaro Tominaga who was the second president.

In this way, Japanese demand for oil increased. Fig.5.60 shows the transition of crude oil supply from 1923 to 1936, in Japan. The total imported crude oil and domestic crude oil reached 1.7 million kL in 1935. Since the crude oil supply in 1997 was 268 million kL, it is only a little over 0.6% of the present age, but it was

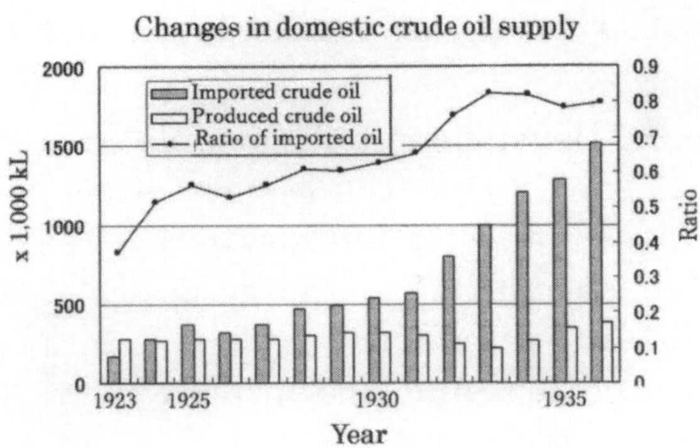

Fig.5.60 Crude oil supply in earlier Showa period (31)

Photo 5.34 Sumida River Oil Tank Station where tank trucks are lined up (around 1935) (31)

still increasing as shown in the figure, and it became that the tank trucks were booming as shown in Photo 5.34. Unfortunately, it is not clear what kind of lorry shipping system was used, but if you imagine from the photo, it will be measured by a level meter.

5.6. History of Variable Area Flowmeter

5.6.1 Development of Variable Area Flowmeter (21)(50)

It is said that the history of the variable area flowmeter dates back to 1868 when Edmond Augustin Chameroy obtained an U.S. patent. Cylindrical and piston area flowmeters were commercialized by New Jersey Meter before 1928. In the 10 years between 1928 and 1938, the one with a built-in indicator for gas by Pittsburgh Equitable Co., the one with a built-in indicator for aircraft fuel lines by Eclipse Aviation Co., the one with a built-in recorder for gas by Lambert Meter Co., and the one with a remote electric type (induction balance principle) by Brown

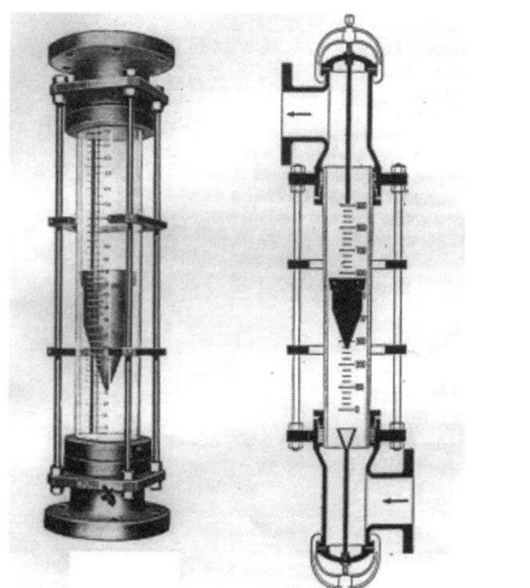

For vertical flow   For horizontal flow

Fig.5.61 Rotor meter of Schutte & Koerting (50)

Instrument Co., had been developed respectively. Early in the last decade, Schutte & Koerting Co., created a free-float type known as a rotor meter, as shown in Figure 5.61. Because of spiral groove cut on the top of a float, the float rotates so that it is located in the center of flow, when fluid flows. It is said that, as a result, the error by friction with wall were reduced. Therefore, its accuracy was good and its flow range was wide enough to allow 1: 200.

The shape was also very simple and tapered glass tube was suitable for clean liquids or gases with normal temperature and pressure.

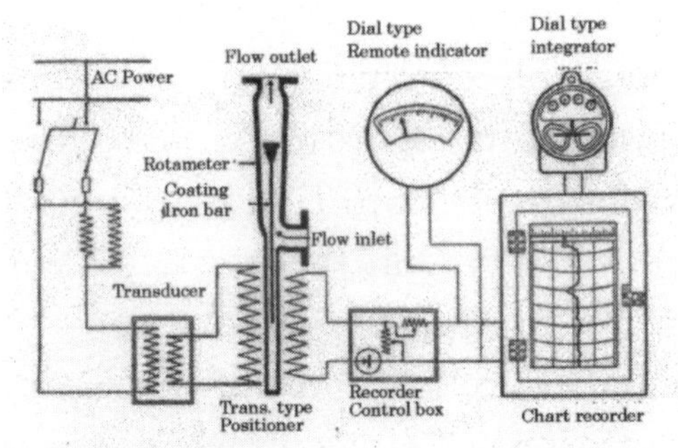

Fig. 5.62 System drawing of Schutte & Koerting (50)

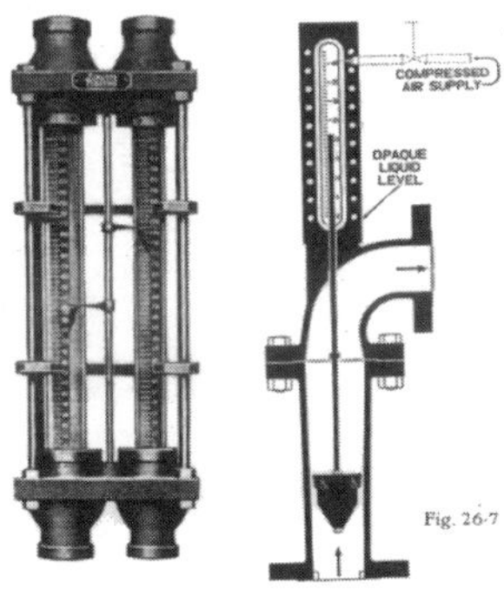

Fig. 5.63 Application example at that time (50)

Fig. 5.62 shows a system drawing concerning the rotor meter which the company equipped with an indicator, recorder, and integrator.

The left side of Fig. 5.63 is an example of this as a twin rotors type meter for fuel-air ratio equipped as standard equipment in General Electric's gas furnace, and the right-side version is for opaque liquid.

From around 1937, rotor meters were mass-produced by Fischer and Porter Co. In 1941, a paper entitled "Reduction of viscosity error in flow measurement with rotameter" (51) was published by K. S. Fisher Blechman and others from Fisher & Porter.

Mr. Blechman was serving as a consultant for Yokogawa Electric Corp. and Oval Corp. for a long time after the war, and had really contributed to the field of fluid measurement in Japan.

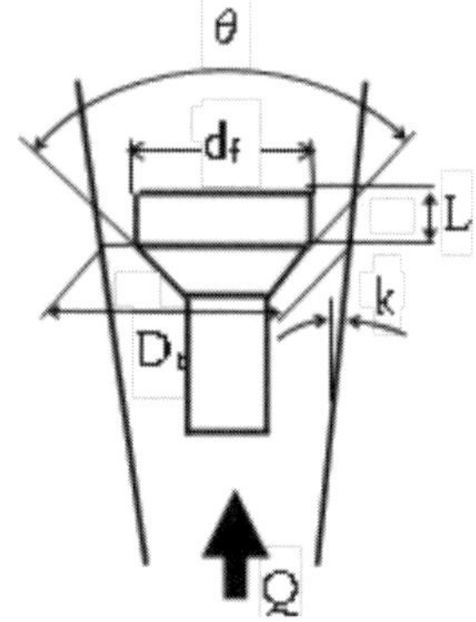

Fig. 5.64 Principle diagram of variable Area flowmeter

The principle of flow measurement regarding variable area flowmeters is well known, but I would like to briefly mentioned this. Fig.5.64 shows a typical tapered pipe type, and the following equation for an area type is expressed by using the theorem of Bernoulli like a differential pressure type. From the pressure difference between the upstream and downstream of float, flow rate Q is;

$$Q = C_1 \frac{\pi}{4}(D_t^2 - d_f^2)\sqrt{\frac{2}{\rho}(P_1 - P_2)} \qquad \text{------------------(13)}$$

Where, P_1, P_2 is respectively, upstream pressure and downstream pressure, ρ: The density of flowing liquid, C_1: Shrinkage coefficient

D_t, d_f: Respectively, inside diameter of cylinder, outside diameter of float in Fig.5.67

5.6.2 History of Other Flowmeters Before World War II

Other flowmeters had not yet been put into practical use before the war (before 1945). However, concerning turbine meters, electromagnetic flowmeters, ultrasonic flowmeters, etc. showed signs of appearing.

1) Turbine Flowmeter

According to R. W. W. Scott, the first turbine flowmeter was developed in the United States in 1938. The schematic diagram is shown in Fig. 5.65, which has spiral blades and was used to measure the flowrate of airlift fuel. Since the life of rocket engines was short at that time, there were no problems even if the bearing life was poor, but subsequent applications required the improvement of bearing characteristics. This was accomplished by two different solutions.

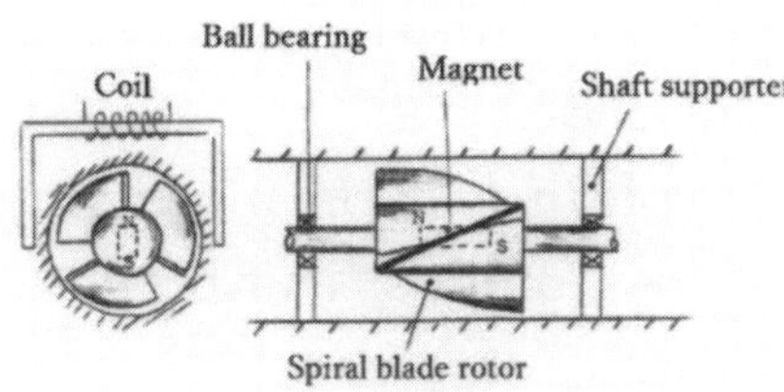

Fig.5.65 Early turbine flowmeter (55)

The first solution was to use a sleeve bearing / thrust sphere combination with a material that can withstand thrust loads and have a tolerable life. As typical examples, tungsten carbide ball bearings and shafts were used for small meters, and stellite sleeves, chrome plating and tungsten balls were used for large meters.

The second solution was to use the force of fluid to reduce bearing friction and thrust loads.

The great thing about this idea was the journal bearing, which created a liquid thin film that was boosted around the axis. This had the effect of reducing the frictional suction torque and of minimizing the thrust load, which was significant for improving performance.

The first of this design was the famous Potter meter shown in Fig.5.66.

To explain a little more why the thrust is reduced, when fluid hits blades, the blades rotate and at the same time, generate thrust force in the direction of flow.

However, as shown in the figure, flow velocity is the fastest because the flow path is narrowest at the part of centrosome (central object) that supports the blades, but pressure is the lowest according to Bernoulli's theorem. Then, since the pressure on downstream side is high, the centrosome is pushed toward the upstream side and a force for reducing the thrust force is generated.

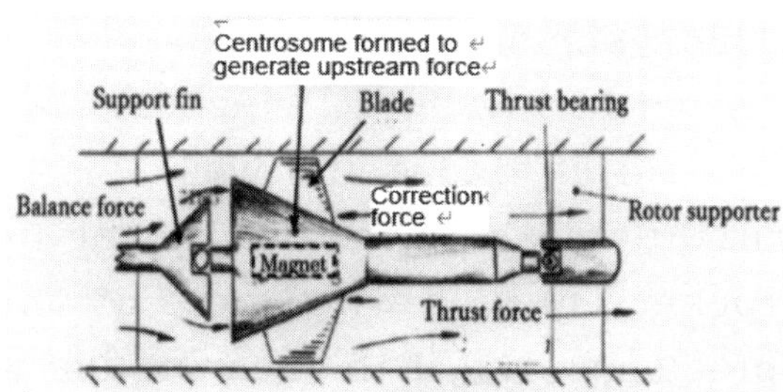

Fig.5.66 Early thrust compensated design (Potter meter) (55)

This design still lives for turbine meters today.

(2) Current Meter

Although it is not for industrial use, there had been current meters for measuring flow velocity and wind speed of rivers for a long time, dating back to the 17th century Hook anemometer. (See Section 2.4.1). In the United States in the 1900s, there were roughly two types for measuring the flow velocity of rivers. The cup type and screw type. The screw types are shown in Fig. 5.67, from the top, Haskell, Hoff, and Ott meter. These meters could measure if the flow velocity was 1 ft / s (0.3 m / s) or higher. The American Geological Survey Society and others had published papers using these current meters since around 1901.

As for anemometer, the Friez remote-indicated type was already used at the airport. [59]

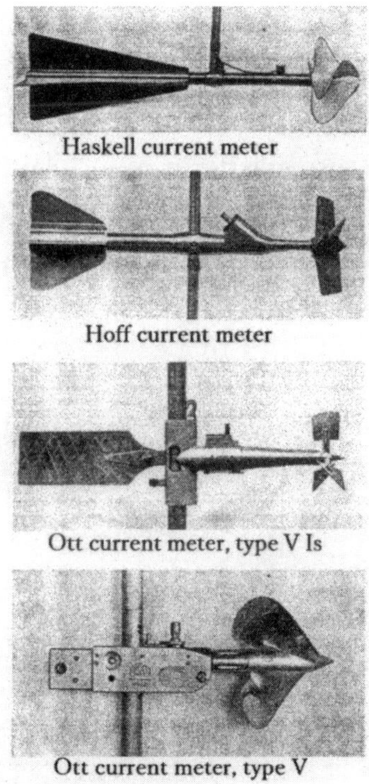

Fig. 5.67 Current meters in 1900s [53]

(3) Electromagnetic Flowmeter [52]

What about electromagnetic flowmeters, which are currently in maximum demand in the industrial field?

As already mentioned, this history dates back to 1832 Faraday's experiments on the Thames. In 1917 Smith and Slepian patented an electromagnet for a ship surveyor. This was to radiate a magnetic field from hull into the water and install electrodes on the hull to measure the speed of the ship.

The first flowmeter device was reported by E.J. William in 1930, but it was only of academic interest.

P. Fabre first used this technique as a device to measure blood flow in 1936. A. Kolin introduced an electromagnetic flowmeter that quantitatively measures blood velocity, but for wider industrial use it would take time until post-war development.

5.6.3 History of Automatic Control and Computers

Current industrial processes are accompanied by temperature, pressure and flow control. Automatic device is now referred to as Automation, but the beginning of Automation dated back to the 18th century, because James Watt designed a centrifugal speed governor to control the speed of a steam engine in 1788. This speed governor is still in use, although it has been slightly modified. Next, in 1858, Isambard K. Brunel invented a steering motor to rotate heavy objects on marine vessels. (54)

However, modern automatic control theory was established quite later, in 1944 when Oldenburg (55) simply introduced the Laplace Transform and developed a theory that emphasized the quality of control. Surprisingly, it was during Hitler's homeland Germany that had been undergoing devastating bombing during World War II.

Therefore, it was necessary to wait after the war for full-scale development.

In 1937, ED S. Smith of the United States described process control in considerable detail at ASME under the title "Industrial Instruments, Their Theory and Application" (57). In this, starting from the theory of automatic control, application examples to temperature control and pressure control were shown.

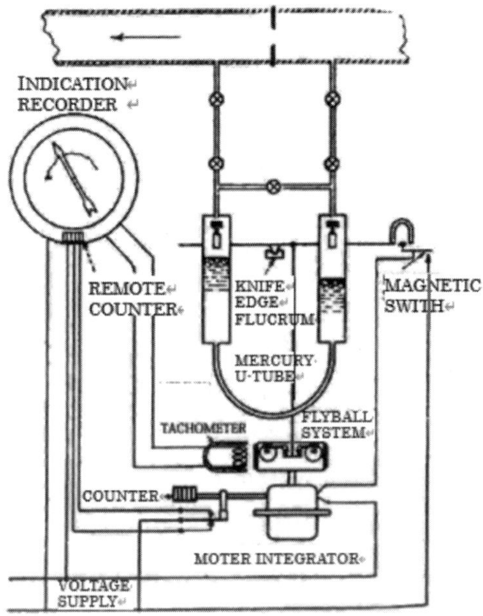

Fig.5.68 Servo operation type zero system mechanical-electric flow meter (57)

Fig.5.68 shows the "Servo-operated zero method of machine-electronics system" applied to flow measurement using an orifice at that time. Let's explain this operating principle in the following order.

(1) As shown in this figure, the amount of mercury entering the left and right sides of a U-tube become unbalanced due

to the differential pressure generated on both sides of an orifice in a pipe.

(2) Then, the U-tube system attempts to rotate counterclockwise with respect to a central knife edge.

(3) At that time, an electromagnetic switch works and electric current flows to rotate a servomotor, accordingly, centrifugal force generated by the rotation of a flyball linked to a servomotor kept the balance.

(4) Flowrate became higher, as the tube tended to tilt more. Accordingly, the number of times of an electromagnetic switch works increased, then, the rotation speed of the servo motor increased. This was the mechanism of automatic control.

(5) A tachometer was a differential device for the rotation of a fly ball and becomes an instantaneous indicator. On the other hand, since there was a substantially quadratic function relationship between head and flowrate, and between centrifugal force and rotational speed, a counter is an integrator that shows the total amount of flow. In addition, this device provides both flow rate and total volume to be transmitted to an indicating recorder.

While investigating flowmeters, the issue of Instrument in the January 1938, "Industrial Fluid Metering and Automatic Control" that R. J. S. Pigott write was noticed. He was the chief engineer of Gulf Research and Development Co., and the chairman of the ASME Research Committee on Fluid meters since 1916, and a heavyweight who was also the vice chairman of ASME. The history of automatic control was glimpsed in this paper. This paper consists of following items.

I. Quantity (volumetric) meter

II. Ratio (inferential type) meter

III. Control

I and II are already mentioned in the previous sections. First of all, in the introduction, he wrote "In the last 10 years (1928-1938), every form of fluid measurement has made considerable progress and its industrial use has doubled. There are two reasons, one is to know exact amount and to calculate the price of fluid (First evaluated where it needs to be calculated, in addition to water supply, electricity and gas). The second is to improve even more the commercial accuracy of the measurement for flowmeters. Furthermore, it cannot be overlooked that the evaluation of superiority derived from the automatic control of industrial fluids has grown in the last decade. That is;

1) The blockage of (a) Weight and (b) Volume
2) Automatic control of flowrate with primary element in response to (a) weight ratio or (b) volume ratio

As is clear from the four classifications here, whether or not the performance of the control system is satisfactory depends largely on the primary measurement element, so that the flowrate measurement has also automatically advanced." In addition, most of the automatic control for industrial fluids were related to temperature and pressure, and it is said that these had developed rapidly in the years 1928 to 1938. As an example of fluid quantity control, a centrifugal tachometer mechanism driven by an Ascania volume meter and flow control by a control valve was given.

Flow-rate control is still far from being generally understood, but it had evolved much more widely in those last 10 to 15 years, at least than in the last quarter century.

It was H.C. Smoot who made a significant contribution to this area. This genius inventor was the first to find that sensitivity, force, speed and stability were fundamental elements of design, to stop guessing and to calculate mathematically. Even after his death, he influenced all types of design for several years.

Around this time Simplex and Builder created control systems for water, filtration and sewage. In addition, the Builder added a Chronoflo transmitter for the operation of controller, enabling remote control over long distances. American Meter commercialized a constant flow controller using a normal differential recorder to operate a rotary valve that receives or discharges compressed air into a thin film type motor control valve.

Ascania was developing a ratio control system shown in Fig. 5.69.

To explain this, as "dominant" flow (1) through an orifice (4) increases, then a diaphragm (9) receives higher differential pressure, as a result, it moves to the left with a jet pipe (13), and oil ejected from a jet pipe enters the left end of a power cylinder (16). Therefore, a valve (3) opens further. As a result, "controlled" flow (2) through an orifice (5) increases, the differential pressure across a diaphragm (10) increases, and it moves to the right, returning the jet pipe (13) to its equilibrium position. The ratio is adjusted by a slider (18) that moves up and down between the jet pipe (13) and an arm (17). Since the two arms of these (17) forms the supporting point at opposite positions, if the slider (18) that is the operating link between the two arms is shifted, it is a mechanism that the load ratio between diaphragms (9) and (10) is changed.

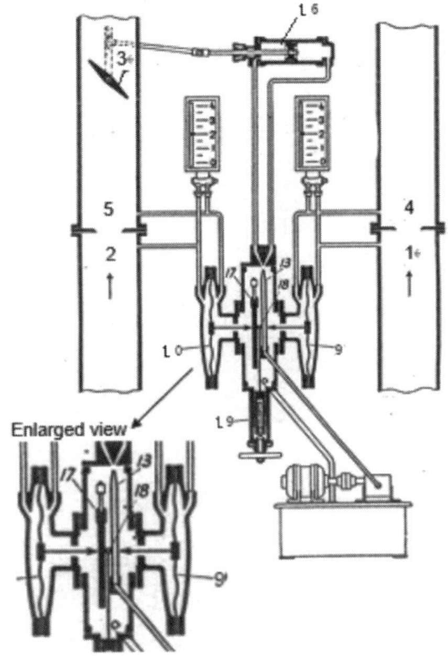

Fig.5.69 Ascania ratio controller (56)

Taylor Co., and Foxborough Co., have each entered flow control fields with their own air activators.
Both companies have also developed practical valve positioners.

In 1937, by ED S. Smith, Jr. et al published "Industrial Instruments, Their Theory and Application" (57) which described Newton's second law as an industrial instrument that measures variables such as flow rate, level, pressure, and velocity. From the above, we develop the characteristics due to inertia and mechanical friction mechanically, introduce the damping coefficient, and discuss the damping characteristics of the vibration system.

In the article mentioned above, it discussed the damping characteristics of the vibration system etc., that measure variables such as flowrate, level, pressure and velocity mechanically which were developed from Newton's second law and so the damping coefficient was introduced. This theory referred to Drysdale's "Electrical Measuring Instruments-Treatment of oscillating rotational systems"(58) 1924, but in the first place, Vibration Engineering would be described by Whittaker in 1917 as "Analytical Dynamics"(59). Furthermore, in 1940, Rudolf Beck published "Thermometric Time Lag"(60), and in that discussion, transient characteristics etc. had already been discussed using the CR equivalent circuit of the electrical system.

The time delay of the thermometer can be traced back to Harper's paper (61) in1912.

Peters published "Experimental Studies of Automatic Control"(62) in 1941, and P. S. Dickey at al. in 1944 announced "Electronic Type Instruments for Industrial Processes"(63).

Among them, Fig. 5.70 is an example of an electronic amplifier for 60Hz AC used for temperature, flowrate, level, etc. at that time. Since it was the vacuum tube era, it was a very large size compared to the current amplifier circuit.

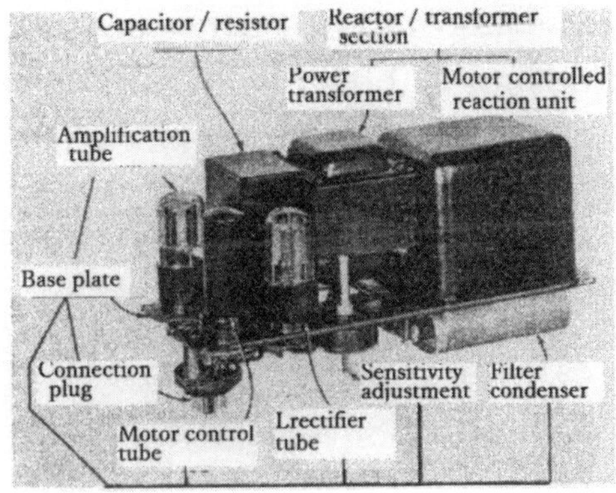

Fig.5.70 Electronic amplification unit at that time (63)

Now, computers are an integral part of current automatic control, so let's take a quick look at the history of computers. The origin of the calculator is the abacus, which was already used in Egypt and China in BC.

The first real calculator that could mechanically add or subtract was invented in 1642 by French scientist Blaise Pascal.

In the same 17th century, Gottfried Wilhelm von Leibnitz, who was German philosopher and mathematician, developed a machine that could perform multiplication, division and square root calculations as well as addition and subtraction.

Photo 5.35 Charles Babbage (64)

In 1812, at Oxford University in the UK, Charles Babbage who was a professor of mathematics designed a computer called the Difference Engine, and 11 years later he devised another calculator called the Analytical Engine. Neither had been commercialized, but it is amazing that the design Babbage was considering had all the basic functions of today's computers shown in Fig.71. Modern computers owe much to his pioneering structure.

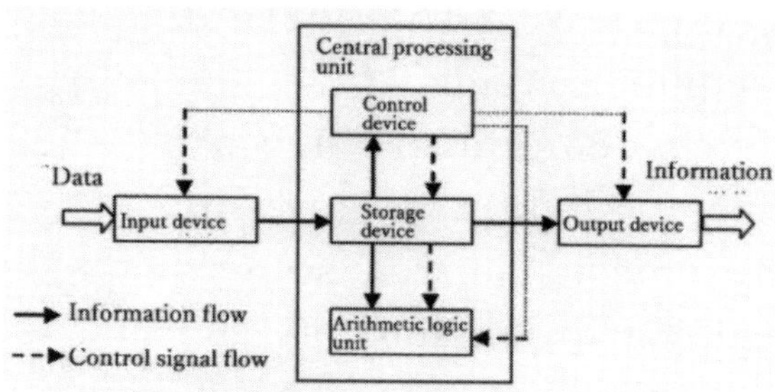

Fig.5.71 Today computer basic functions (64)

The key-operated calculator invented by Felt in 1885 was the first successful commercial adder. A few years later, W. S. Burroughs produced cash registers, and revolutionized the commercial industry. Today, few stores will not use a cash register that can calculate accounts and issue receipts. Fig.5.72 shows these historical computers.

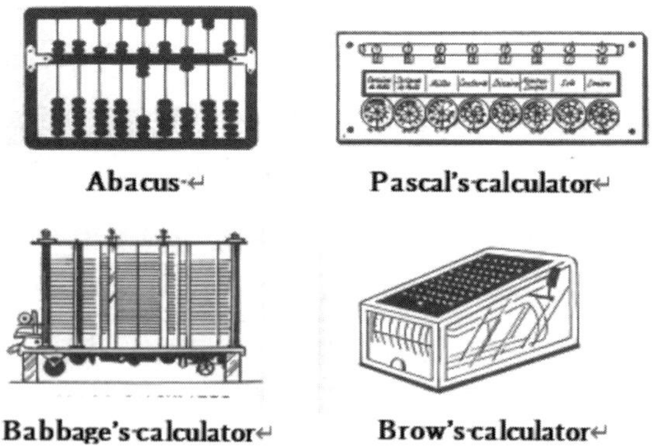

Fig.5.72 A number of historical calculators (64)

In 1887 Hermann Hollerith used the punch card method. It was during World War II that electrical data processing demonstrated true character.

A very important ballistic calculation was needed for newly developed weapons. An analog computer developed by Vannevar Bush was commonly used. The computers of this type had been successfully produced at numerous universities under government contracts.

However, in some cases, computers were needed that could perform faster and more accurate calculations than analog computers, then digital computers had been developed.

Dr. H. H. Aiken of Harvard University designed an automatic continuous control computer or a semi-electric machine called the Harvard Mark I for the Weapons Navy Arsenal. The Mark I was built in collaboration with International Commercial Machinery Company (IBM) and was completed in 1944.

Meanwhile, Dr. J. W. Mauchly first thought of the all-electrical type computer for general purposes in 1943 at the Moor Institute of Technology in Pennsylvania. This device was called the ENIAC Computer (Electric Numerical Integrator and Calculator), and it was completed in 1945.

Photo 5.36 is a part of ENIAC, and it is considered to be first counting type electronic computer in the world. From the left, it shows the panels, of start-up unit, of divider / square root calculator, of accumulator, and of two high-speed multipliers respectively. Approximately 18,800 vacuum tubes, 1500 relays, 70,000 resistors and 6,000 switches were used at ENIAC. However, the calculation speed was about 1000 times faster than the Mark I, so they said, " ENIAC has made the Mark I look as old as the water wheel." ENIAC 's capabilities were epoch-making at that time, but the failure rate was high and the performance was unstable because many vacuum tubes were used for the arithmetic element. It was also a trend of the times that vacuum tubes (first-generation computers) eventually disappeared and transistors (second-generation computers) replaced them.

Photo 5.36 First counting computer in the world (65)

Another pioneer was Dr. John von Neumann.
He contributed to a computer called the "Whirlwind", which was built by the National Standards Bureau as SEAC (Standards Electronic Automatic Computer). He was also one of inspirations for the built-in program system. In this way, it can be said that digital computers have just begun just before the end of the war. However, it was not difficult to imagine that the combination of automatic control technology and computer technology would lead to a dramatic leap forward for the industry association.

With the above, it seems that the "history of flow measurement" before the war (before 1945) has been almost completely investigated.

Now, let's move on to the "History of Flow Measurement After World War II", as the worldwide postwar reconstruction using automation and computers had begun, and flow measurement would make a great leap forward in line with that trend

References

(1) "100 People of Modern Water Supply", 100 People of Modern Water Supply Selection Committee Edition, Nihon Suido Shimbun Co.

(2) "Measure and Protect for 50 years", Japan Gas Meter Industry Association

(3) Mitsutomo Yuasa "Chronology of Science and Culture", Chikuma Shobo

(4) Walter C. Kerr "The Mechanical Equipment of the New South Station, Boston, Mass" Transactions of ASME, vol.21,1899

(5) Tokyo city water works, Engineering vol.7,1901

(6) Yutaka Hosoi "Flow Dynamics (1)" Tokyo Denki University Press

(7) Takashi Nakata "Dr. Kalman" Oval, Specialty N0.8, 1990

(8) Koichi Yumoto "Illustration: Encyclopedia of Meiji's Origin Things" Kashiwa Shobo

(9) Toshihiko Miyake " Old Photo Book of Railway " New Person Traffic Company

(10) John T. Wilkin "Rotary Pumps" Transaction of ASME, vol.24, 1902

(11) W. B. Gregory "The Pitot Tube" Transaction of ASME, vol. 25, 1903

(12) 12 Thos. R. Weymouth "Measurement of Natural Gas" Transaction of ASME, vol. 34, 1913

(13) Francis P. Fisher "Establishing a Standard of Measurement for Natural Gas in Large Quantity" Transaction of ASME, vo1. 38, 1916

(14) Robert F. Farhart "Deviation of Natural Gas from Boyle's law" Transaction of ASME. vol.38, 1916

(15) "Report of the Committee on Standardization of Flanges and Pipe Fittings" Transaction of ASME,vol.40, 1918

(16) *Kesakatsu Koizumi, "To Fluid Specialists, Message from Past: Episode in Development Process of Flowmeter", Oval Specialty, No.4, 1989.1.*
(17) *R. S. Medlock "The Historical Development of Flow Metering" Brown Boveri Kent Plc.*
(18) *Tadashi Tatsuda, "Measurement of Pressure and Handling of Measuring Instruments" Science and Industry, Vol. 1, No. 6, 1925*
(19) *Kesakatsu Koizumi, "To Fluid Specialists, Message from the Past: Episode in the Process of Developing a Flowmeter", Oval Specialty, No.2, 1989.1.*
(20) *L.K. Spink "Principle and Practice of Flow Meter Engineering" The Foxboro Co.,1930*
(21) *"Flow Measurement A to Z", edited by Japan Measuring Instruments Federation*
(22) *Shozo Matsushiro, "Revised Flowrate", Corona Publishing Co., Ltd.*
(23) *ED S. Smith, JR.et al "Industrial Instrument, Their Theory and Application" Transaction of ASME, 1937*
(24) *"McMillan World Science History Encyclopedia 2", Hara Shobo*
(25) *R. J.S. Pigott "Industrial Fluid Metering and Automatic Contro1" Instrument January 1938*
(26) *M. F. Behar "Flow Measurement and Control" Instrument, August 1932*
(27) *Instrument, January 1950*
(28) *Claude B. Nolte "History of Bellows-actuated Flow Meter" Instrument, January 1950*
(29) *American Meter Co. "Inducted into Head Service" The Oil and Gas Journal, vol.39, No.22, 1940*
(30) *The Oil and Gas Journal, vol.39, N0.16, 1940*
(31) *"100 Years History of Nippon Oil" Nippon Oil Co., Ltd., 1988*
(32) *Charles Singer "History of Technology 9" Chikuma Shobo*
(33) *"Oil Handbook 2000" edited by Nisseki Mitsubishi Corporation*
(34) *M. F. Behar "Flow Measurement and Control Chapter 23, Positive Meters" Instrument, December,1932*
(35) *M. F. Behar "Flow Measurement and Control Chapter 23, Positive Meters" Instrument, January,1933*
(36) *Edger E. Ambrosius "volumetric-Type Fluid Meters" The Oil and gas journal.vol.39, N0.13, 1940*
(37) *M. F. Behar "Flow Measurement and Control Chapter 23, Positive Meters" Instrument, February 1933*
(38) *M. F. Behar "Flow Measurement and Control Chapter 23, Positive Meters" Instrument, April 1933*
(39) *Joachim Böhler "Handbuch der Wassermessung" R. Oldenbourg Verlag, München. Wine 1965*

(40) E. W. Jacobson "Some Fundamenta1 Consideration in the Design and Application of Displacement Meter" The Oil and Gas Journal, 1940, June 20
(41) E. W. Jacobson "Calibration of Displacement Meters on Volatile-Liquid - Petroleum Fraction" Transaction of ASME.1941, Nov.
(42) R. F. McCormick "Positive-displacement Metering of Liquids" Instrument, 1953, July
(43) Paul Renfrew "Liquid Displacement Meters" Instruments, 1952, Sept.
(44) Yoji Katsuragi "History of the Rise and Fall of Western-USA-Japanese Automobile Manufacturers" Grand Prix Publishing, 2004.8.28
(45) McMillan: "Historical Statistics of North and South America", Toyo Shorin
(46) McMillan: "Historical Statistics of Asia, Africa and Oceania" Toyo Shorin
(47) Nobuo Iitaka: "Evolution of Gasoline Sales Equipment (1), (2)", Sekiyu Jihou,1945. July
(48) "Fifty Years with the Oil Flow (1911-1961)", edited by Tokyo Tatsuno Seisakusho Co., Ltd.
(49) "100 Years' History of Tominaga ", edited by Tominaga Seisakusho Co., Ltd.
(50) M. F. Behar "Flow Measurement and Control" Instrument, 1933 Dec.
(51) Kermit Fisher Seymour Blechman "Elimination of the viscosity Error in Flow
Measurement with Rotameters" Instruments 1941 March
(52) R. W. W. Scott "Developments in Flow Measurement" Applied Science Publishers
(53) Floyd A. Nagler "Use of Current Meters for Precise Measurement of Flow" Transactions of ASME 1933
(54) Popular Science" vol.9,10,Grolier
(55) R. C. Oldenbourg, H. Sartorius "Dynamik selbsttaiger Regelungen" Oldenbourg Muchen 1944
(56) R. J. S. Pigott "Industrial Fluid Metering and Automatic Control Instrument,1938 Jan.
(57) ED S. Smith "Industrial Instruments, Their Theory and Application" Transaction of ASME 1937
(58) M V. Drysdale "Electrical Measuring Instrument-Treatment of Oscillating Rotational Systems" Ernst Benn, Ltd. London 1924
(59) E. T. Whittaker "Analytical Dynamics" 1917
(60) Rudolf Beck "Thermometric Time Lag" Transaction of ASME 1940
(61) D. R. Harper "Thermometric Time Lag" 3rd, U. S. Bureau of Standards, Bulletin N0.185, 1912, p.659
(62) J. C. Peters "Experimental Studies of Automatic Control" Transaction of ASME 1941

(63) *P. S. Dickey et al, "Electronic-Type Instruments for Industrial Processes" Transaction of ASME 1944*
(64) *"Poplar Science" vol.9 and 10*
(65) *Translated by Ryosuke Watanabe; "History of Digital Computational Tools", Just systems, 1995*

6. The Birth of Flow Measurement from the Postwar Reconstruction

6.1 From Postwar to Modern Times

6.1.1 History of Postwar Flow Measurement (1)(2)(3)

In the Second World War, Axis powers such as Japan, Germany, and Italy were completely attacked by Allied Forces. The Pacific War ended with the Potsdam Declaration in Japan on August 15, 1945. I would like to focus on the world postwar history of flow measurement centering on Japan.

Photo 6.1 MacArthur arrived in Japan (2)

Japan had been under the control of GHQ (General Headquarters) for some time. Photo 6.1 is of Douglas MacArthur, the Supreme Commander of the Allied Powers, who arrived in Japan on August 30, 1945.

Looking back, Japan entered the Pacific War on December 8, 1941, with a shocking attack on Pearl Harbor. The Japanese army consecutively won the opening battles and captured the US military base in the Philippines. General MacArthur, commander of the United States Army in the Far East, escaped from Leyte Island in the Philippines, departing with the famous phrase "I shall return." Eventually, the Allied Forces dominated the war and the MacArthur regained the Philippines on October 20, 1944, as the Commander-in-Chief of the South West Pacific Area, as per his words he left behind. The US military captured Japanese troops one after another in the Marshall Islands, Saipan Island, and Okinawa, and he landed at Atsugi Airfield as the Supreme Commander of the Allied Powers. (1)

Now, it is confusing about how to approach the history of flow measurement after the war, since the industry had expanded so rapidly after that, it seems to be very uncontrollable.

However, as a prologue, I would like to borrow the words of Mr. Mitutomo Yuasa, "The discovery of fire is estimated to be a million years ago, and it must be said that it was an epoch-making event in the development of human culture along with the invention of language and letters, but the discovery of second fire (nuclear energy) will have the same significance." (3)

Unfortunately, two atomic bombs exploded in Hiroshima and Nagasaki in Japan, leading to the unconditional acceptance of the Potsdam Declaration, but with peaceful use, it could be said that nuclear power would replace coal, oil and natural gas as future energy sources.

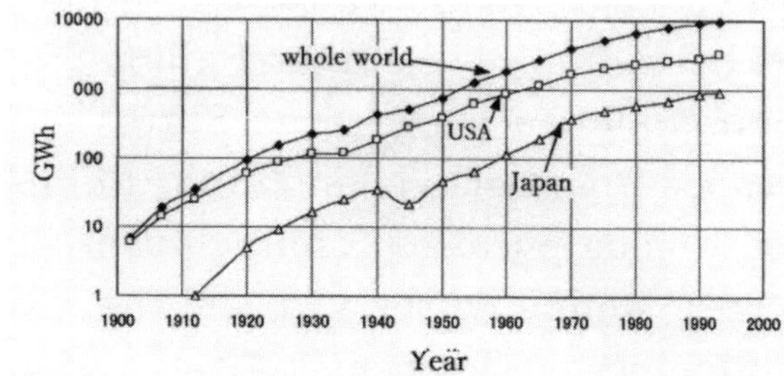

Fig.6.1 Change of Electric Energy (4)

The degree of global economic growth can be seen by looking at the change of electricity energy. Fig.6.1 was created by me based on McMillan's statistical data (4). As shown, the electric generation that started in the United States in 1902 is about 500 GWh of electric energy on the whole world scale in 1945, when World War II ended, and the United States accounts for 53% (265 GWh) of that.

Japan falls to the level of the 1930s before the war because of the destruction by air strikes etc., and Japan's suffering can be read. Looking at modern 1993, the world's power generation is 10,000 GWh, which is 20 times that at the end of the war, 12 times that of the United States, but about 40 times that of Japan.

The United States accounts for 34% of the world, and Japan also had increased economic power, accounting for about 10%. This is an amazing reconstruction from the post-war rubble. However, the share of nuclear power in the world's primary energy consumption had not shown such an explosive growth. After all, according to safety issues, we must see that nuclear power generation is sluggish, and when will the "second fire" become the main force? The steam engine was invented in the 18th century, but the "century of steam" was in the 19th century, and the generator was invented in the 19th century, but the "century of electricity" was in the 20th century. Then, since nuclear theory was discovered in the 20th century, does history tell us that the "century of nuclear energy" is one century behind in the 21st century?

6.1.2. Reconstruction from Rubble (6)

Mark Gain, an American newspaper reporter who landed at Atsugi Airfield on December 5, three months after the end of the war, said: "As we approached Yokohama, the seriousness of Japan's damage became clearer. It was a ruin as far as the eye could see. People in tattered Kimonos seemed distraught... I couldn't find any new building. Skeletons of locomotives and railroad vehicles were left untouched on the rails... The many bodies of buses and automobiles looked like the pulling out of their intestines were rolling on the side of road. It was exactly like a human-made desert there. Everything was hard to see and rough, and it was hazy from the dust rising from the collapsed bricks and stucco."

I also returned from Manchuria Dalian in February 1947, and was packed into a crowded train from Sasebo, and witnessed the devastated cityscapes of Hiroshima, Osaka, and Yokohama from the train's window. Also, when seeing grocery shoppers behaving violently in the train, it made me feel sick of my childhood.

As you know, plutocracies were dismantled by the command of GHQ. However, some special and intense measures were needed to get the Japanese economy out of paralysis and restarted. On December 24, 1946, the Cabinet decided and implemented an epoch-making "Gradient Production Method" that concentrated all measures on coal and steel. Thus, postwar reconstruction began with the coal and steel industries.

What about the oil industry related to flow measurement? The bombing of Japan's oil facilities by the US Air Force, which began in June 1945, was concentrated in refineries. Only local refineries such as Showa Shell Sekiyu's Nankai, Nippon Oil's Niigata, and Kashiwazaki survived the bombing.

The response of the compensation delegation to the almost scrapped refineries were initially strict, and they even said, "Japanese refinery facilities are non-modern and not subject to compensation, so they should be scrapped.". But the Draper Survey Team said in the report, "Japan's oil refineries are small, but the technical level is reasonable, and if they are revived and put into operation, it will save 10 million dollars a year." So, the story was completely changed.

American private companies such as "Standard," "Shell," and "Caltex" boarded in the GHQ Div.4 as an oil advisory group.

These companies created JOSCO (Japan Oil Storage Co.), a company for the purpose of oil storage, and started sales activities to foreigners in August 1948. Oil industries were also in contact with foreign capital from early on. In February 1949, Tonen and Standard Vacuum, Nisseki and Caltex, and Mitsubishi and Tidewater signed foreign capital alliance agreements in March respectively. Since then, partnerships had been formed one after another with Showa Shell (from June) and Koa Caltex (from July), and the reconstruction of the Japanese oil industry was begun.

There is a secret story of a tearful oil company right after the end of the war. “Oil companies in post-war struggled to maintain their operations. Not only were refineries banned, but they were even banned from doing anything on the premises. That is why Nippon Oil made medicines, pomades, and shoe shavings outside the premises to connect their jobs. This was because the remaining oil was used. Maruzen Oil and Mitsubishi Oil were also among them.

What was strange was the Toa Fuel Industry's salt making and ice making and Idemitsu Kosan's radio repair business. Tonen planned to switch to ammonium sulfate production, but did not obtain permission from the GHQ authorities. The radio repair business of Idemitsu used its own large area and directly managed retail organization.”

Photo 6.2 Photo 2 Nippon Oil tank truck with Caltex mark immediately after the war [2]

Photo 6.2 shows a group of Nippon Oil tank trucks affiliated with Caltex, lined up at the Sumida River oil depot immediately after the war. Looking at the refueling landscape, it seems that flowmeters were used for shipping.

6.1.3 Dawn of Japanese Measurement Technology [7]

In this way, in Japan, measures for the coal and steel industries had begun by the " Gradient Production Method ", and the latest US technologies had been introduced to the oil industry as well. As introduced in Section 5.4.1, orifice meters and positive displacement flowmeters were already used abundantly in refineries in the 1940s in the United States. As automatic controls and digital computers began to be used, these latest technologies were also required of the steel industry and refineries.

For this reason, in addition to Toshiba, Hitachi, Mitsubishi, etc., which were comprehensive technology companies, old and new companies such as Yokogawa Electric Works,

Yamatake, Hokushin Electric Works, Tokico, and Oval for automatic control and flow measurement were beginning to move. As the interest of measurement in each industry increased, the first issue of "Keisoku (Measurement)" was published by the Measurement Round-table Conference on December 25, 1951. The publication, entitled "Measurement for Industrial Reconstruction," by Jiro Yamauchi, a professor at the University of Tokyo and chairman of this council, stated as follows.

"This time, we have decided to publish the magazine "KEISOKU (Measurement)", and its purpose is to try to connect people are gathered from various fields related to measurement, such as manufacturers and users of measuring instruments, people involved in certification and inspection, and researchers related to measurement, and that by cooperating introduce the current state of measurement inside and outside the country, convey correct usages, solve various measurement problems, and mainly aim to improve the technology of measurement sites at each factory, finally contributing to the development of Japanese industry.

------ When the urgent industrial recovery of Japan after the end of the war began, the first thing that all the specialist engineers of ammonium sulfate industry and the steel industry who came from the United States pointed out was the wind of neglecting measurement in Japan. In addition to improving production capacity, they explained how important measurement is for lowering basic units and rational use of fuel.

------In order to improve this situation as soon as possible, both manufacturers and users need some kind of institution to cooperate with each other to supply goods and durable measuring instruments and to use them properly. ----- The momentum for "measurement" to be born from these demands have matured. (Omitted thereafter)"

In this first issue, the chairman, Jiro Yamauchi, wrote directly about "standard measuring instruments for the steel industry".

An example is shown in Fig. 6.2, and it was recommended to use the one shown in Table 6.1, which shows the flowmeter.

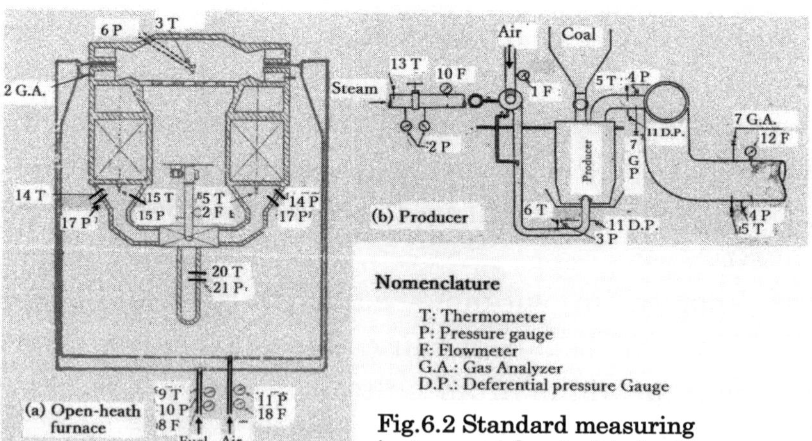

Fig.6.2 Standard measuring instrument for steel industry

In "Measurement" Vol.1 No.2, an article introducing the products of Tokico R type flowmeter and Oval gear type flowmeter was published, and at the end of the book, advertisements for flowmeter of both companies (Tokiko Kiki Kogyo Co. and Oval Kiki Kogyo Co.) were first posted as shown in Fig. 6.3.

Table 6.1 Flowmeter used for steel process

Process	No.	Measuring fluid	Detector	Instrument
Open-heath furnace	2	Air flowrate	Orifice plate, Pitot tube	Sinking bell type or Ring balance type
	8	Fuel oil flowrate		PD meter, Area meter or Deferential pressure type
	18	Primary air flowrate		Area meter
Producer	1	Air Flowrate	Orifice plate, Pitot tube	Sinking bell type or Ring balance type
	10	Steam flowrate		
	12	Gas flowrate	Orifice plate	

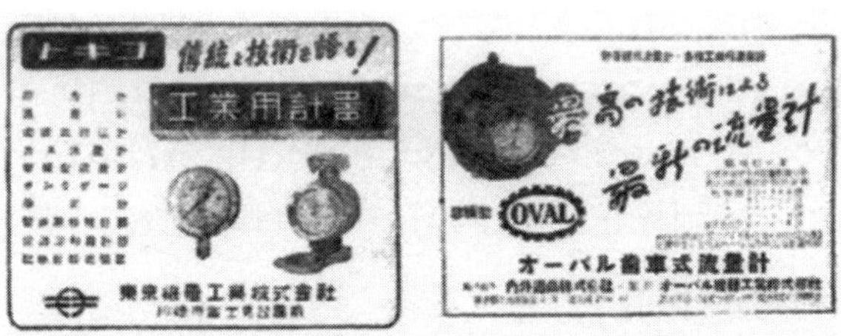

Fig.6.3 Advertisement of Japanese flowmeters

6.1.4. Biography Series of the Founders in Instrument Making

Here, I would be like to summarize the founding stories of several instrument makers that contributed to society in the postwar measurement field, excerpted from the company histories.

(1) Yamatake [8]

Photo 6.3 Takehiko Yamaguchi

Takehiko Yamaguchi was born on July 2, 1869 (Meiji 2) in Nishida-cho, in the castle Kojima, Satsuma Domain. The Yamaguchi family approached the Shimazu family as a senior samurai of the Satsuma domain, and his grandfather served as a side servant of domain lord. In 1887 (Meiji 20) he entered the Department of Mechanical Engineering, Tokyo Vocational School (currently Tokyo Institute of Technology), and after graduating, he had worked for Ministry of Agriculture and Commerce. Photo 6.3 shows a Portrait of Takehiko Yamaguchi.

Korekiyo Takahashi (who became a Japan prime minister in 1921) persuaded Zenjiro Yasuda (the founder of Yasuda Financial) to produce domestic nails around 1896, and recommended sending Takehiko to the United States as a research engineer.

But Takehiko was warned by a nail factory in the United States, and had hard times conducting technical investigations. He finally placed an order for 10 wire drawing blocks and 100 nail machines, and returned to Japan in 1897, having visited Europe after his job.

Yasuda built the Fukagawa Nail Factory and succeeded in domestically producing nails in 1898, but due to a series of hardships, the raw material for nails did not come in from the United States, and the factory caught fire because of the Great Kanto Earthquake in 1923, so the factory was demolished.

He worked for Yasuda Kogyo, but decided to become independent in 1906 and established the agency "Yamaguchi Shokai" for the Forva Vogt Co. Immediately afterwards, he renamed it the "Yamatake Shokai (firm)" by using "Yama" of Yamaguchi and "take" of Takehiko and became the sole agent in Japan for Schuchardt & Schütte of Germany. Machine industry grew dramatically with the Russo-Japanese War, leading to an increase in imported machinery.

The assembly of industrial instruments was started in 1932, and in the following year, Yamatake Keiki Seisakusho (Yamatake Instrument factory) opened. From 1916, it imported and sold the industrial instruments of Brown Co., Ltd. of the United States, and in 1933, it started production of inductance bridge flowmeters and float type liquid level gauges.

In 1939, Nippon Brown Keiki Co., Ltd. was established, and in 1941, the company name was changed to Yamatake Instruments. In 1949 after the war, Braun was absorbed by Honeywell, but in 1956, it deepened the tie-up with Honeywell and changed its name to Yamatake Honeywell Instruments. It is well known that Yamatake Honeywell has been active in the measurement industry since then. Takehiko lived a long life of 93 years.

(2) Yokogawa Electric (9)

Tamisuke Yokogawa studied architecture at Imperial University and opened an architectural office in 1890 (Meiji 23), and in 1902 (Meiji 35) he opened a "Yokogawa Public Office" in Nihonbashi, which supervised architectural design. This is the origin of Yokogawa Electric. Around 1914, Ichiro Yokokawa and Shin Aoki thought that electrical measuring instruments would be needed in Japan in the future, so they visited and travelled around Europe and the United States. After returning to Japan, on September 1, 1915, they launched the signboard of the "Electrical Instrument Research Institute" as a small barracks factory next to the Yokogawa Chemical Research Institute. Initially there were only 12 members. An initial goal was to domestically produce integrated wattmeters, but researched the reduction of production costs for mass production, and at the same time focused on research and the trial

Photo 6.4 Portrait of Tamisuke Yokogawa

Photo 6.5 First President Ichiro Yokokawa (left) Managing Director Shin Aoki (right)

production of indicated electric meters.

In the spring of 1917, they succeeded in producing prototypes such as an ammeter, voltmeter, and wattmeter for the first time. Photo 6.6 shows an initial electric meter for a switchboard. In regards to flowmeters, they started selling in 1934 (Showa 9), but it was from electromagnetic flowmeter in 1962 (Showa 37) that it began in earnest. In 1955, it signed a 10-year technical assistance contract with Foxborough for industrial instruments and automatic regulators. In 1963, it formed also, a technical tie-up with Hewlett-Packard Company.

Photo 6.5 Initial electric instrument

(3) Hokushin Electric Works [10]

Syohei Shimizu was born in December 1893 (Meiji 26) to a farmer in Sutu Village, Fuji District, Shizuoka Prefecture.

Shimizu had excellent grades when he was in an elementary school, but his parents did not allow him to go on to junior high school because his

Photo 6.7 Portrait of Syohei Shimizu

family was poor, so when he was 16 years old, he left home with savings of 83 yen.

He attended a night school at Tokyo Physics School, graduated with a second grade in 1918, and got a job at an electrical laboratory, but stopped after one year and served as an assistant to Hantaro Nagaoka famous for research into magnetism for four years, and seemed to have caught his eye.

There, he came into contact with experimental machines and instruments and learned for himself how the Japanese precision machinery industry was lagging behind.

At that time, pyrometers were imported, so he decided to domestically produce them. Nagaoka took care of this and in 1918 (Taisho 7), he put up the signboard of "Hokushin Electric Works" in a corner of Katsumata Electric Co.

It is said that the Analects of Confucius state,

"Governing by the power of virtue can be compared to the Pole Star (Hokushin)*, which remains fixed in place while all the other stars orbit respectfully around it."*

He chose Hokushin for his company name, because he wanted to be like the polestar. The Pyrometer was difficult to delivery, but the first product

Photo 6.8 Japan's first domestic pyrometer

was completed in February 1920 (Taisho 9).
Suspended thermoelectric temperature indicator model 20 is shown in Photo 6.8.
Then it is said that Shimizu discarded all the parts of General Electric meters such as ammeters and voltmeters, which had been a side business until then and placed the company's luck on the pyrometer.

During the war, it manufactured aviation weapons, but lost 85% of its facilities in the bombing, dismissed all 2,500 employees in 1945, after which it hired 450 again and resumed making industrial instruments.

After that, he formed a tie-up with Bendix in the United States in 1956 and introduced liquid flow meters, etc., and in succession with Burton and Fisher & Porta, expanded the differential pressure mechanism and industrial instruments, and also entering in the field of automation.
It had a good track record.

However, in 1983, it suddenly merged with Yokogawa Electric Mfg. Co., Ltd. and became Yokogawa Hokushin Electric Co., Ltd., but when the company name changed to Yokogawa Electric Co., Ltd. in 1986, the name of Hokushin virtually disappeared.

(4) Tokico (11)

Tokico's predecessor was Tokyo Gas Electric Engineering Co., Ltd., which was founded in 1910 (Meiji 43).

After the "February 26 Incident" in 1945, the military initiative in Japan became stronger, and then the wartime regime rapidly shifted.

As the military control over the industry was strengthened, military demand priority became more apparent in the Gas Electric, and it was concluded that it was the best policy to make the instrument department independent for development purposes.

Photo 6.9 Portrait of Hajime Hasida

Then, Tokyo Kiki Kogyo Co., Ltd. was founded on July 19, 1937. It produced parts for aircraft and automatic vehicles, pressure gauges, vacuum gauges, gas flowmeters, orifice meters, etc.

In particular, pressure gauges have been in full-scale production since 1918 at the Gas Electric Engineering Company, and in February 1945, just before the end of the war, they set a monthly production record of 5,645 units based on a command to increase production from the military.

It is no exaggeration to say that Hajime Hashida, the managing director, had been taking the measure of all the products, from the instruments and other vehicle equipment that followed, to the aircraft vaporizers that boasted the best in Japan during the war.

Since its founding, he had helped successive presidents, including watches and scales even after the war, and in 1953, he had made many achievements such as introducing the technology of oil injection pipe type automatic control device of Ascania into West Germany mentioned in section 5.6.3. In 1949, he succeeded in advancing a gasoline weighing machine and established Tokico Oil Equipment Co., Ltd. In addition, he developed the roots meter shown in Photo 6.10 and succeeded in receiving a large order for 150 aircraft fuel (JP-4) flowmeters (150m^3 / h) from the US Far East Air Force stationed in Japan in March 1951.

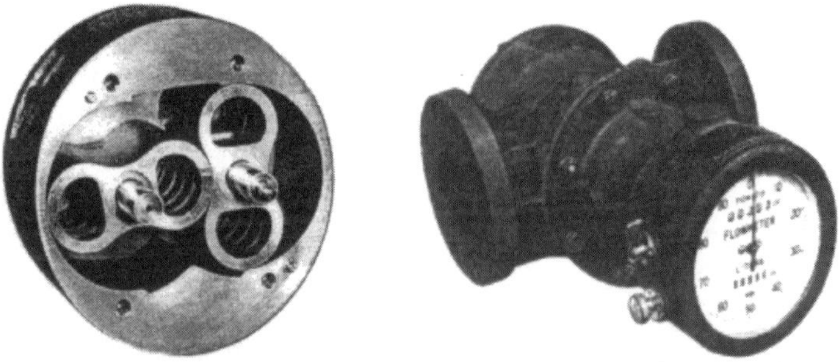

Photo 6.10 Roots meter

Unfortunately, he passed away at the age of 65. He made great
achievements in the field of flow measurement. However, Tokico Co., Ltd., itself merged with Hitachi Ltd., on October 1, 2004 by focusing on automobile-related businesses. Tokico Techno Co., Ltd. and Tokico Giken Co., Ltd. currently exist as affiliated companies.

(5) Oval (12)

Photo 6.11 Portrait of Jun Kashima

Jun Kashima was born on September 1, 1912 (Meiji 41) in Wada, Minouchi-mura, Saeki-gun, Hiroshima Prefecture. His father was engaged in agriculture and forestry, but he was 50 years old and Jun wasn't very strong, so when he was six, All the farms, forests and houses were sold and the families moved to Hiroshima.

Kashima was always in the top class in elementary school. He then went to Syudo Junior High School, next to Hiroshima High School and was active in a speech club. On the verge of graduating from high school, the principal recommended him to Inoue Kaoru's breathtaking Inoue Scholarship Society, and he became acquainted with Gisuke Ayukawa, the first president of Nissan

(Note: Inoue Kaoru was one of the heavyweights of the Meiji government.)

He went to Tokyo Imperial University to become a lawyer in the future. His senior said, "I think it's difficult for a man like you to get ahead, so it will better not to be a bureaucrat." So, Kashima decided to enter the business world and joined Tokyo Gas Electric Engineering Company, which was in the process of being rebuilt in 1933. He did his job well, and his skill was recognized by Nao Uchiyama, the managing director, and he took on a heavy responsibility at Tokyo Motor Co., Ltd. and later transferred to Hino Heavy

Industries. However, the military intervened in the company too and the wartime system became severe. Because of opposition to the army, he suddenly received call-up papers on September 19, 1943 (31years old), and went to North China to wander through the living and dead on the battlefield. Fortunately, however, he returned to Japan on New Year's Day 1946 after the end of the war.

In 1946, he was asked by the Nao Uchiyama (the president of Sonoike at that time) to join Sonoike, and became the manager of the Tsujido factory. When he visited Shousuke Murakami who worked for Nissan Automobile Co., who he became acquainted with during the Tokyo Automobile Industry era, "Unfortunately I couldn't succeed with my oval gear needed for use in automobile engines with my own hands, but could you bring it to the world?" It is said that President Uchiyama also agreed with the realization of the story. However, his benefactor, President Uchiyama, died suddenly, and in October 1947 Kashima left Sonoike for various reasons.

After that he seriously considered commercialization with Tamotsu Yamamoto, who was originally from the Arsenal Department at the University of Tokyo and worked at Sonoike, and as a result, it was decided to apply the oval gear to a low-speed engine instead of a high-speed engine. Kashima suddenly remembered that Murakami had said "There was a company in Germany that made flowmeters using oval gears". *(Author's note: it was Bopp und Reuters)*

However, no matter how much he searched, he could not find a user in Japan, and there was no patent. So, he decided to develop it independently. However, in February 1949, Shosuke Murakami died suddenly too, and Kashima quietly renewed his determination to Murakami's will. When he told Gisuke Ayukawa about the commercialization of the Oval Gear, he was strongly opposed to saying, "Because Murakami couldn't do it for the rest of his life, I won't say bad things, you should stop it" However, Kashima did not change, and finally on May 10, 1949, he established the Oval Gear Engineering Co., Ltd. with 10 comrades. Up to this point, it was a turbulent life. After that, as shown in Photo 6.12, the prototype of the oval gear flowmeter was finally successful, and it came to be used in many fields such as petroleum, petrochemical, electric power, and gases etc.

[12] **Photo 6.12 Prototype No. 1 of Oval meter**

It is very impressive that many of the founders of measuring instrument companies received considerable support from the tremendous politicians, managers or scholars of the time. It is also very interesting that Tokico and Oval, which have been competitors in flow measurement, were connected at Tokyo Gas Electric Engineering Company once upon a time.

Internal structure of Murakami special pump

Oval gear for rotary engine

Photo 6.13 Murakami type special pump and oval gear for engine (13)

6.2 Flowmeter Development Front in Japan after the World War.

6.2.1 Oval Meter and Roots Meter

Differential pressure type flowmeters were used in Japan before the war, but it was the Oval Meter and Roots Meter th at appeared extensively in postwar Japan.

With postwar reconstruction, oil, chemical, electric power, and gas plants had been built one after another, and moreover the reconstruction of industry accelerated, especially because of the special dem and caused by the outbreak of Korean War in June 1950. Under such circumstances, in process automation, Yokogawa Ltd. And Electric Works Hokushin Ltd. became competitors with each other, and in the flowmeters used in plants, Oval Gear Engineering.,Co., Ltd. and Tokyo Engineering Co Ltd. (Tokico) did also.

As mentioned in the previous section, Oval was founded in 1949, but its oval gear was a technology inherited from Shosuke Murakami of Tokyo Automobile Industry Co., Ltd. (The predecessor of Nissan Motor Co., Ltd).

In Photo 6.13, the upper one shows the internal structure of the Murakami special pump using three-leaf gears, and the lower one shows the trial product of oval gears for rotating internal combustion engines by Nissan Automobile Co., Ltd. But no oval gear engine was realized. r

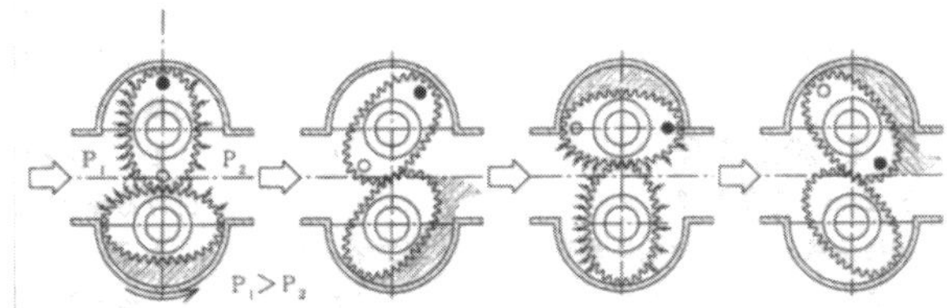

Fig.6.4 Principle diagram of oval meter

Oval independently developed a flowmeter based on the principle shown in Fig.6.4 and succeeded in the first prototype flowmeter for hot water shown in Photo 6.14 in July 1949. Its initial performance is shown in Fig. 6.5. Incidentally, in order to sell flowmeters completed in this way, employees expanded their business by the so-called "Entotsu Commercial method", which searched for Entotsu

Photo 6.14 The first prototype of a flow meter for hot water (14)

(chimney) with their motto "There is a boiler where there is a flowmeter will be necessary where there is a boiler." chimney. After that, riding on the wave of enforcement of New Measurement Law, deliveries to chemical companies, oil industries and electric power complexes increased.
Photo 6.15 shows a group of oval flowmeters delivered to the Toa Fuel Industry Shimizu Plant around 1960. (About 10 years later).
On the other hand, Tokico started to develop roots meter used as a blower as a liquid measuring instrument, but initially it did not perform well, and after many difficulties i n designing and

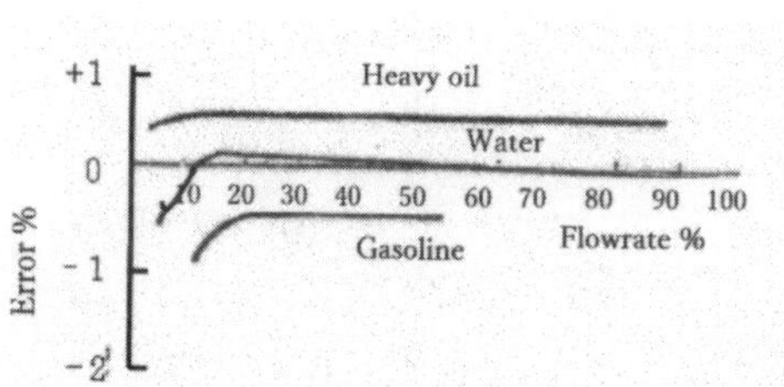

Fig. 6.5 Instrumental error characteristics of early oval meters (14)

Photo 6.15 Oval flowmeters group (15)

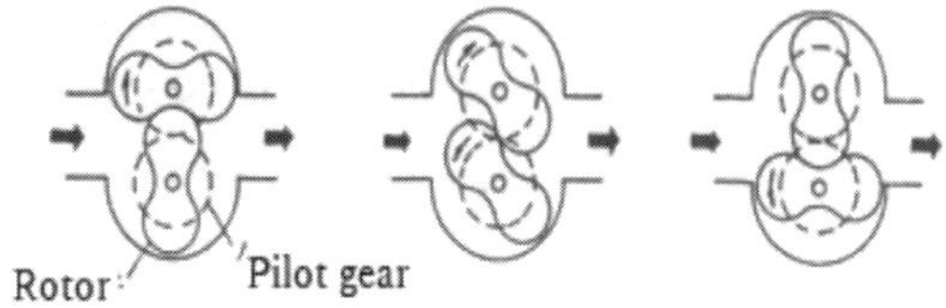

Fig.6.6 Principle diagram of roots flowmeter

Photo 6.16 Tokico R type flowmeter (14)

manufacturing it was finally completed as an industrial high-precision flowmeter in August 1950. Although the operating principle is shown in Fig. 6.8, its roots can be said to be two oval gears with two teeth. However, as shown in the figure, the pair of rotors will not rotate without the pilot gear. Photo 6.16 is the original Tokico R-type flowmeter. *(Sorry old blurred photo.)*

Fig.6.7 shows the initial performance curve (Size:2

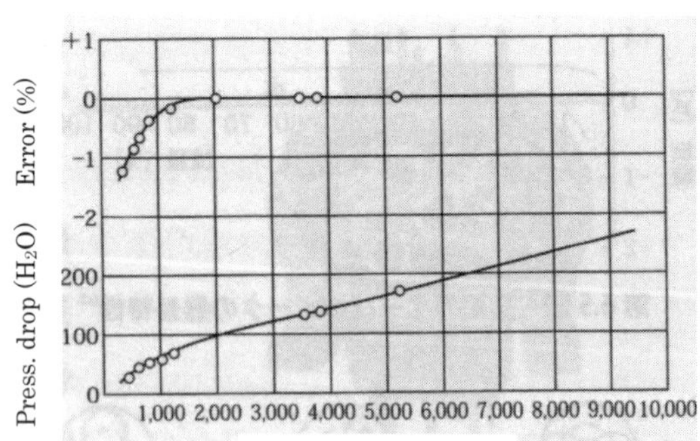

Fig.6.7 Performance of Tokico R flowmeter

inches) and Table 6.2 is the standard product list of roots flowmeter.
Tokico also entered the gasoline measuring instrument in 1950, competing with Tatsuno and Tominaga. Photo 6.17 is the first gasoline measuring instrument installed at the Izumi pressure booster of the Tokyo Metropolitan Waterworks Bureau.
Oval completed the model change of oval flow meters in 1965,
which systematized a pocketless and compact oval meter

Table 6.2 Standard product list (14)

Nominal tube Diameter (Inch)	Min. Flowrate (m^3/h)	Max. Flowrate (m^3/h)	Normal Flowrate (m^3/h)
1	350	9,000	1,000～7,000
11/2	350	9,000	1,000～7,000
2	1,200	12,000	1,500～10,000
3	3,000	30,000	4,000～15,000
4	6,500	65,000	10,000～50,000
5	12,000	120,000	20,000～75,000

with a magnet coupling method. In addition, it licensed the oval flowmeter, etc. to Brooks Instruments Inc. of the United States, and in 1966 signed a mutual technical assistance contract with Bopp unt Reuters of Germany to strengthen its technical capabilities.
Both the oval meter and the roots meter have good performance, but the rotation of rotors

makes a periodic non-constant velocity motion.

Photo.6.17 first gasoline measuring instrument

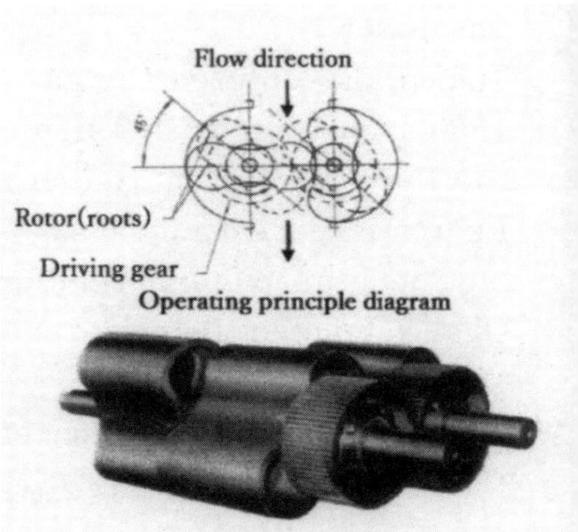

Fig. 6.8 Non-fluctuation Roots Meter

Tokico decided to deliver a roots meter with a high-frequency transmitter to Showa Shell Sekiyu, but pulse intervals were uneven due to the periodic non-constant velocity motion of rotors. To solve this problem, a non-fluctuation roots meter was developed in 1964. As shown in Fig.6.8, the rotor that was out of phase by 45 degrees resulted in perfect constant speed rotation.

As a result, an epoch-making volumetric flowmeter that brings low noise and low vibration to a large meter by Tokico appeared!

Please pay attention here to the Murakami special pump shown in Photo 6.13 (P-237). The Murakami special pump shown in Photo. 6.13, has the same non-fluctuation roots and was a combination of three-leaf rotors that were 60 degrees out of phase. It is considered that this would be also aimed at constant velocity rotation, low noise, and low vibration.

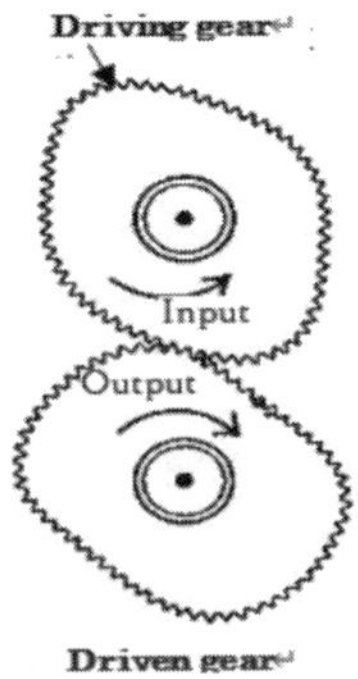

Fig.6.9 Constant speed gear (18)

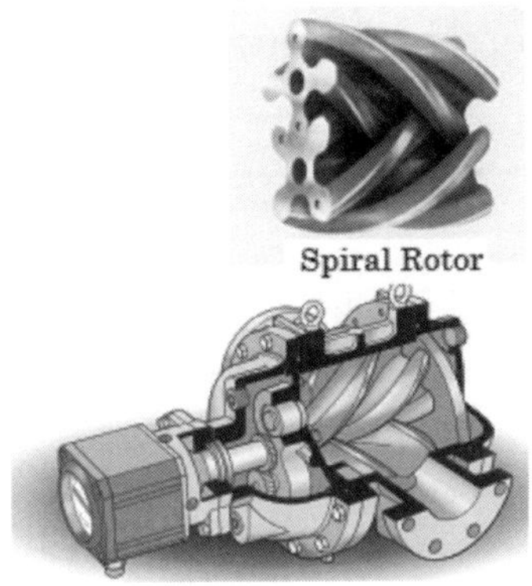

Fig. 6.10 Uniflow Oval (UF II) (18)

The solution that happened to the oval meter for this non-constant speed rotation was the invention of the constant speed gear by a gifted engineer, Koji Hotta. Instead of making the speed constant with the rotor, after taking out the rotation of rotors, the constant speed gear shown in Fig.6.9 was engaged to make the output rotation at a completely constant speed.

Furthermore, in 1971, Oval commercialized the Uniflow Oval (UFII), which rotates at a completely constant speed. This is a helical gear type, and its construction is shown in Fig. 6.10.

This method was also been commercialized by Tatsuno. Mechatronics and Ryutai Kougyo later

The companies that handle volumetric flowmeters as of 1995 are classified in Table 6.3. (All Names are abbreviation.) Table 6.3 Companies handling volumetric flowmeters in Japan (19)

Flowmeter model name	Company name
Oval type (Gas, Liq.)	Aichi , Oval, Kinmon, Tokico, Yuainics, Ryutai Kogyo,
Helical gear type (Liquid)	Oval, Tatsuno, Yuainics, Yuei, Ryutai Kogyo,
Roots type (Gas, Liquid)	Kinmon, Schlumberger, Tokico, Dresser Japan,
Rotary piston type (Liq.)	Ono Sokki, Kinmon, Schlumberger, Dresser Japan, Nitto Seiko,
Rotary vane type (Gas)	Tokyo Flowmeter Research Institute, Dresser Japan, Nitto Seiko,
Reciprocating piston type	Tatsuno,
Membrane type (Gas)	Aichi, Kansai gas meter Kinmon, Schlumberger, Takenaka
Wet gas meter	Kansai gas meter, Kinmon, Schlumberger, Takenaka
Disk type (Liquid)	Schlumberger
Rotation ball type	Teikoku Denki

Please refer to Section 5.5 about the system of each type.

6.2.2 History of Electromagnetic Flowmeter

As the history of electromagnetic flowmeters is detailed in Kazuto Kenno's paper (20), it will be described based on this. The electromagnetic flowmeter is based on a principle that an

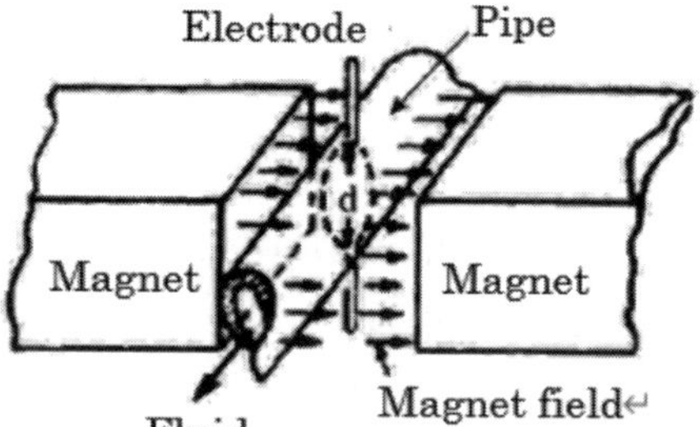

Fig.6.11 Principle diagram by Colin (20)

electromotive force is generated in a conductor that move across a magnetic field.
Fig. 6 .11 shows the principle diagram.

An electromotive force (E) proportional to the average velocity v (cm / sec) of fluid is obtained in an electrode tube provided in a direction perpendicular to the magnetic field and the flow of fluid.

The electromotive force E is expressed by the following Equation.

$$E = B \cdot d \cdot v \times 10^{-8}(Volt) \text{-------------------------------} (14)$$

Here, B is magnetic flux density (Gauss), and d is the inner diameter of tube (cm).

This equation (13) is derived by A. Kolin in 1945.

In 1952 James, S. Arnold derived the same result with this equation for a rectangular cross section.

Photo 6.18 Experimental electromagnetic flow meter

In 1952, H.G. Elrod et al. published a paper (21) relating to this. So, let's elaborate on it a little. Elrod et al. developed an electromagnetic flowmeter to measure the flow of molten corrosive metal through a small diameter tube.

Photo 6.18 is the flowmeter. The tube used polystyrene and advanced alloy, the inside diameter is about 5 mm. Both tubes are non-magnetic.
Experiments were done by using the equipment shown in Fig. 6.12 and mercury is used instead of alternative solutions. The results of the experiments are shown in Fig. 6.13. After that, J.S. Arnold, A.J.

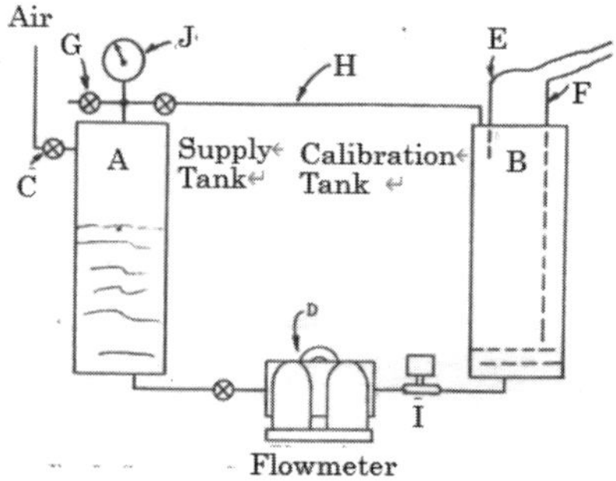

Fig.6.12 Experimental device (21)

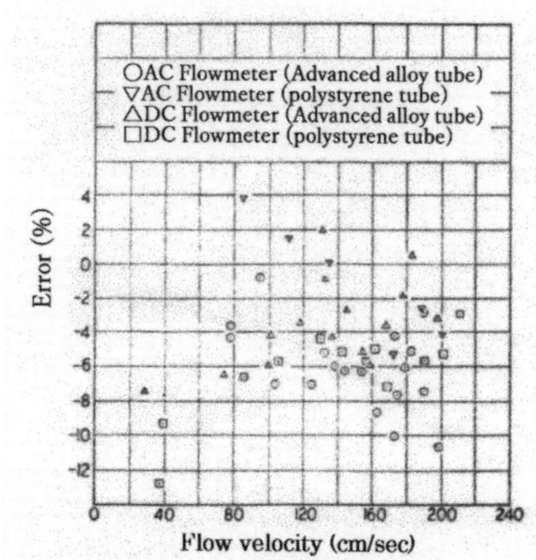

Fig.6.13 Error% of experimental results from theoretical formulas (21)

Morris, L.M. Grossman and others published papers, and in Japan in 1953, university graduation papers such as the University of Tokyo etc. were scattered, but these were not yet put into practical use.

"Magnaflow" based on the research of Eugene Mittelmann in 1951 was the first to be commercialized in the world.

On the Instruments Exhibition at the 1954 International Conference, the two companies Foxborough and Hauser announced products that had high accuracy as an electron tube equilibrium meter and enabled automatic control of the flowrate. In addition, Tobi Co. of the Netherlands gave the product name Tobi Flux.

GE had commercialized electromagnetic flowmeters that measured the flowrate of molten metal that was thought to be used in nuclear boilers.

On the other hand, in Japan, research had been conducted at the University of Tokyo, Mechanical Testing Laboratory, Hitachi, and Hokushin Electric Works since around 1950, and it attracted attention when Hokushin Electric Works announced a prototype of an electron tube type balance meter at the 1955 Spring Meeting of Japan Society of Applied Physics.

In addition, Kaken made the world's first successful marine magnetic flowmeter using the earth's magnetic field, which had obtained good results. It can be said that it was a brilliant thing that realized the idea of Faraday. (22)

Photo 6.19 Japan's first electromagnetic flowmeter by Hokushin (23)

Hokushin Electric Works completed Japan's first electromagnetic flowmeter shown in

Photo 6.19 in 1957. According to an engineer who used to work for Hokushin Electric Works in the past, the first model FL280 had a KelF resin coating on the inner surface of the electromagnetic flowmeter, but the Teflon resin sleeve in 1961 was put into practical use and reliability significantly increased.

Furthermore, in 1964, the FL282, the world's largest ultra-large-diameter electromagnetic flowmeter with an inner diameter of 2300 mm, shown in Photo 6.20 was delivered to the Tokyo Asaka Water Purification Plant. (If compared to

Photo 6.20 The world's largest Hokushin electromagnetic flowmeter (23)

the woman standing at the entrance of the flow meter, you can see its very large size)

This structure is the one in which a wound iron core and a split winding coil are attached to the inner surface of steel pipe and centrifugally cast with polyurethane resin.

Yokogawa also had a technical tie-up with Foxborough to manufacture large-diameter electromagnetic flowmeters. When a magnetic field is applied, although coil must be wound so that the magnetic field is uniform in the cross section with the electrodes with respect to the flow of fluid, all of Foxborough, Yokogawa Electric, and Hokushin Electric adopted winding method shown in Fig.6.14 (b).

In addition, the winding method (a) is a method, patented by the Hokushin Electric Works at that time, where the shape of flowmeter can be compact. (24) Since then, the practical growth of electromagnetic flowmeters is well known, and applications had diversified, except for petroleum and oil which are non-conducting fluids.

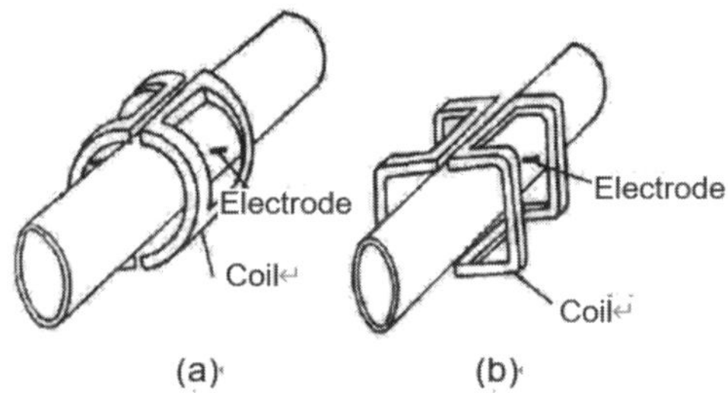

Fig.6.14 Method of wind the electromagnetic coil (24)

In 1974, Vincent Cushing announced that it could measure when the measuring section pipeline was conductive, or even if the conductivity of the fluid was very small compared to that of conduit. So, it was suggested that the flowrate of petroleum could also be measured. (25)

However, it seems that there will be still many problems so an electromagnetic flowmeter that can measure petroleum has not been commercialized at present.

Looking at the excitation method, the one using commercial frequency became widespread in the 1960s, and there was a major change in the introduction of the square wave excitation method from the 1980s. In addition, the field of measuring fluids has expanded according to improvements in lining technology, and the advent of ceramic linings, in particular, triggered the expansion of applications in food, chemistry, and sewerage.

In the 1990s, it was extremely difficult to measure the flowrate of low-conductivity fluids such as pure water, sugar solution, and alcohols in the field of electromagnetic flowmeters, but after the capacitance detection method with high-frequency excitation was adopted, stable measurement became possible.

Photo 6.21 Two-wire Electromagnetic flowmeter (19)

In addition, the 2-wire type electromagnetic flowmeter with the output of 4 to 20mA (not necessary of local power supply) appeared. Photo 6.21 shows a recent 2-wire electromagnetic flowmeter. Since the electromagnetic flowmeter has no components inside the conduit, pressure loss is almost zero and power consumption is low, it is just a demon with a gold rod. It was showing rapid demand growth in the last 10 years. In 2001, it finally overtook the differential pressure flowmeter among the flowmeter models and jumped to the top of the domestic production value in Japan.

Since it is also applicable to liquids other than petroleum, if an electromagnetic flowmeter that can really measure petroleum appears in the near future, it will pose a threat to the volumetric flowmeter market. Therefore, as of 1995, the main companies handling electromagnetic flowmeters in Japan were Aichi Tokei Denki, Emerson, Oval, Kinmon Seisakusho, Sakura Endless, Tokyo Instrumentation, Toshiba, Tokimec, Nitto Seiko, Nippon Flowcel, Hitachi, and Fuji Electric, Yamatake, Yokogawa Electric, and Ryutai Kogyo, as many as 15 companies in Japan.

6.2.3 History of Ultrasonic Flowmeters

The history of ultrasonic flow measurement [26] is said to date back to 1919 when W. Kunze wrote a German patent for wind speed measurement methods. The German patent for O. Rutten in 1928 and the US patent for C. Chilowsky in 1932 also describes the flowrate measurement method using

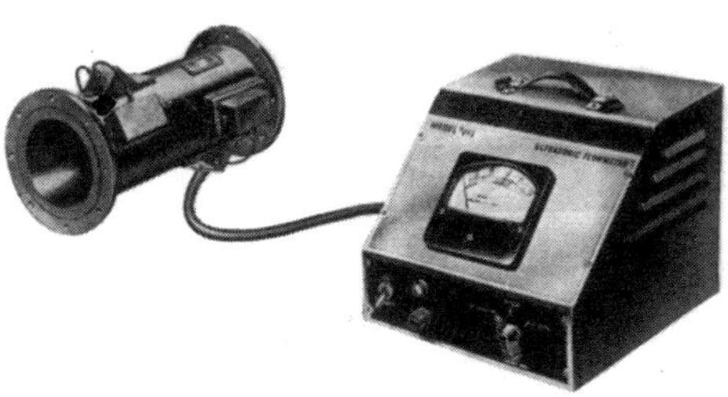

Fig.6.15 The first ultrasonic flowmeter by Maxon [27]

ultrasonic waves, but they were still incomplete and did not reach practical stage.

From 1942 to 1943, there was an invention for a ship speedometer using the Doppler method.

In 1949, the speed of sound in air was measured by a single-around method, and in 1950, the measurement of the amount of water in a large duct appeared.

In 1953, Kalmus announced a design of transmitter and a test equipment suitable for measuring blood flow in a small tube.

In 1955, a product of ultrasonic flowmeter for liquid flow measurement as shown in Fig. 6.16 of by Jack Kritz of Maxon Co. first appeared. Concerning this development, Fisher & Porter also supported.

As shown in Fig. 6.16, it is consisted of two ultrasonic transmitters and two ultrasonic receivers. Volume flowrate, mass flowrate, and fluid density could be measured using a 10 MHz crystal.

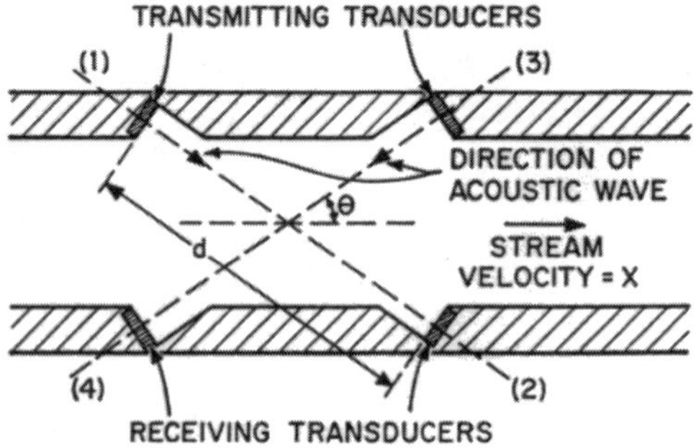

Fig.6.16 **Structure of ultrasonic flowmeter**

It is said that flowrate range was 20: 1, accuracy was within 2% of full scale, and linearity was within 1% of full scale.

Fig.6.17 shows the propagation method of continuous wave. Assuming that the speed of sound of ultrasonic waves is v and fluid velocity is x in Fig. 6.17, the arrival time T_1 of sound waves from (1) to (2) and the arrival time T_2 of sound waves from (3) to (4) are expressed by following two equations.

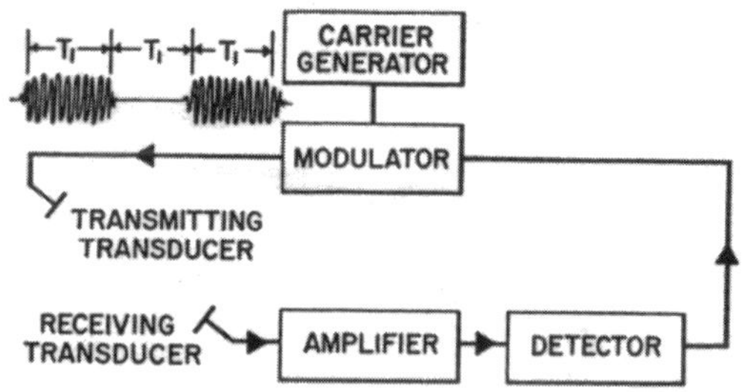

Fig.6.17 Continuous wave propagation method (27)

Here, θ is the incident angle of ultrasonic wave to the flow of the fluid stream.

$$T_1 = d/(v + x\cos\theta) \text{------------------------------}(15)$$

$$T_2 = d/(v - x\cos\theta) \text{------------------------------}(16)$$

From these two equations, the flow velocity v is explained as

$$v = \frac{d}{2\cos\theta}\left(\frac{1}{T_1} - \frac{1}{T_2}\right) = \frac{d \cdot \Delta f}{2\cos\theta} \text{--------------}(17)$$

If Frequency difference Δf is measured, Flow velocity (v) can be obtained. This method is currently called “Sing-around

Technique”.

In this paper, it was also pointed out that most important error of ultrasonic flowmeter is caused by velocity distribution in a tube.

The ultrasonic flowmeters of Maxon Co., were put into practical use as the flowmeters of aircraft fuel. Since then, successful application examples had been published in numerous literatures.

In Japan, in 1957, Niwa et al. presented "Simultaneous transmission / reception type ultrasonic flowmeter by switching electron tubes", which was a phase difference method, at the Institute of Electrical Engineers of Japan. In 1964, Tokyo Keiki Co., Ltd. created a practical measuring instrument for ultrasonic flowmeters for large-diameter pipes, and after numerous experiments, it was used to measure the efficiency of water turbines at hydroelectric power plants. Eventually, all power companies began using it as a portable flowmeter. (28)

After that, various developments and improvements progressed, and the measurement principle methods of ultrasonic flowmeters are now roughly classified as follows (28)

(1) Propagating time difference method (time difference method, correlation method, sing-around method, phase difference method)

(2) Doppler method (Doppler method, pulse Doppler method)

(3) Cross-correlation method

Also, if these were classified by structures, the following 2 types were roughly divided,

(1) Clamp-on type (A type installed outside piping)

(2) Measuring tube type (Sensor insertion type).

The Doppler method assumes that the moving speed of floating solids contained in fluid is equal to flow velocity, and sequentially emits ultrasonic waves in the fluid, and receives reflected ultrasonic waves from the floating solids, and finds the flow velocity from the frequency changes of the received wave. Currently, there are 20 companies that handle ultrasonic flowmeters as members of the Japan Measuring Equipment Industry Association. It is probably the reason why the ultrasonic flowmeters have potential in the near future.

Under such circumstances, Fig.6.18 is a clamp-on type ultrasonic flowmeter for gas measurement, which had been considered impossible until that time, and was the world's first one developed by Panametrics Co., Ltd. I was curious about what kind of technology was used to make it possible.

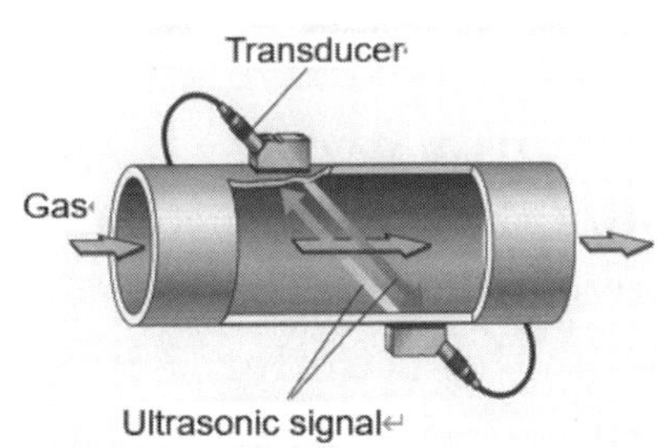

Fig.6.18 Clamp-on type ultrasonic flowmeter for gases

After that, it was devised so that it was not affected by the flow velocity distribution. Fig. 6.19 shows this, and Fig. (a) reflects it twice on the pipe wall, and Fig. (b) reflects it four times to reduce the influence of the flow velocity distribution. For this reason, the accuracy of measurement was good, and it became widely used for the transactional measurement of natural gas.

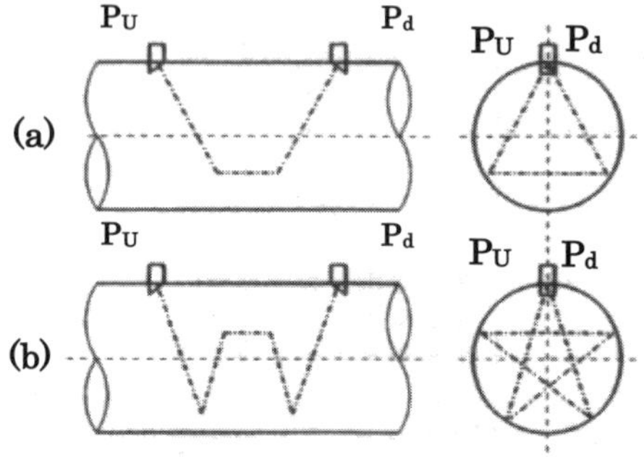

Fig. 6.19 Reduction method of flow velocity distribution effect by multipath (29)

According to the ARG Advisor Group (Wil Chin) Global Market Research on Ultrasonic Flowmeters, the ultrasonic flowmeter market was nearly $406 million in 2002 and would span the next five years, it would grow at an annual growth rate of 7.9% and was expected to reach just under $600 million in 2007 (see Fig.6.20).

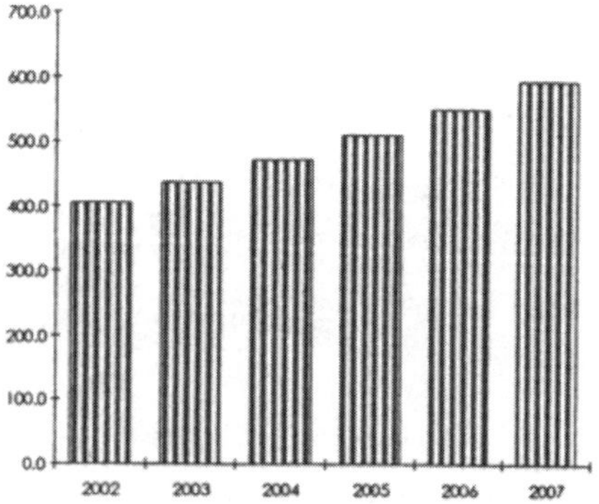

Fig. 6.20 Global market research for Ultrasonic flowmeters (Million dollars) (31)

Ultrasonic flow meters were adopted for the transaction flow rate measurement of liquid petroleum refined products and natural gas etc., and grew steadily in this market.

In draft standards for liquid hydrocarbon trading flow applications by the American Petroleum Association (API), natural gas trading applications by the American Gas Association (AGA), international trading flow applications by the International Legal Metering Authority (OIML), etc., were said to support the adoption of high-precision ultrasonic flowmeter. By region, shipments to Europe, the Middle East and Africa led other regions. (31)

6.2.4 History of the Vortex Flowmeter

(1) About Dr. Karman Again

Photo 6.22 Dr. Karman (center) Photo taken when he came to Japan (32)

The Vortex flowmeter is a flowmeter using the Karman vortex street discovered by Theodor Airplane and Birth von Karman, as described in 5.1.6 "Invention of the of Modern Fluid Dynamics"

Dr. Karman had visited Japan before the second war. I would like to tell you about an interesting episode during this.

President Kawanishi of Kawanishi Aircraft (now Shin Maiwa Industries, Ltd.) asked Dr. Karman to "Please come to my company for 400 yen a month" to create a wind tunnel for airplane experiments. When Kalman consulted with his mother, she said, "Don't go to a barbaric country like Japan. Please refuse." Kalman was in trouble and he thought that President Kawanishi would give up, if he demanded double salary. So, he wrote a letter saying, "If it's 800 yen a month -----". Speaking of 800 yen at that time, it was the price to build a house. Kalman had no choice but to come to Japan because President Kawanishi replied, "800 yen is fine."

Karman was also the first person to commended for his aerodynamic achievements when J.F. Kennedy created the Order of Science in the United States.

Photo 6.22 was taken when he came to Japan, but I wanted to know when Dr. Kalman came to Japan, and so I it looked up.

At the beginning of "Theory of Flight" (33) translated by Ichiro Tani for the Japanese version published in 1955, it states that "Dr. Kalman spent several months in Japan nearly 30 years ago." In addition, Ichiro Tani's "Translator's Words" states that he was invited to lectures in the United States, Japan, China, India, etc. from 1926 to the following year. From this, it is probable that he came to Kawanishi Aircraft in 1926 (Showa 1). In addition, when facing the camera, on the left of Dr. Karman, the person wearing Japanese clothes is probably President Kawanishi.

Photo 6.23 One of the vortex row photographs in Kalman's paper (34)

Dr. Karman was also said to be German, but was born on May 11, 1881 in Budapest, Hungary.

Photo 6.23 is one of the Karman vortex series photographs published in Dr. Karman's dissertation on the Karman vortex (34).

It is surprising that flow visualization photography technology was already established 94 years ago.

By the way, in the morning of December 12, 2019, a huge Karman vortex appeared side by side over the sea south of Korean Jeju Island, South Korea. Kyushu in Japan is reflected in the upper right. The image was taken by NASA's Earth observation satellite Terra.

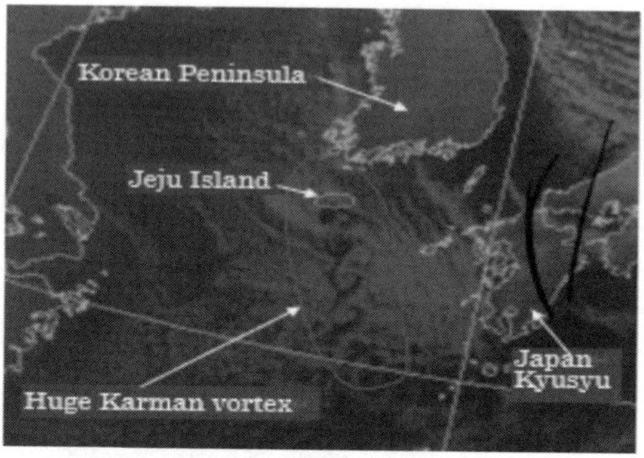

Photo 6.23-1 Huge Karman vortex appeared by the Jeju island ,2019,12,12 from Inter Net

(Author's check: If assuming the width of the vortex row is 80km and the speed of wind is 30m, the cycle will be around 4 hours by using the Karman vortex equation. That is, it will move one pitch after about 4 hours.)

(2) Appearance of the Industrial Vortex Flowmeter

The Karman vortex flowmeter was actually developed in 1968, five years after Dr. Karman died.

The following episode is a story heard by Akira Takata, who later became the president of Oval Corporation.

D. F. White with the cooperation of A. E. Rodely of the United States created a laboratory in the basement of his house, where he created wind with a small blower, detected Karman vortices coming out of the vortex generator with two The waveform seen on the oscilloscope was highly fluctuating and unstable. He was wondering how to improve the waveform while thinking and making mistakes every day, but his wife started complaining because his research expenses had run out and he was spending his living expenses, so he was considering quitting research. When he casually held his hand over the vortex generator, he felt that the waveform of the oscilloscope suddenly. Improved.

He wondered "what's happening?" So, held his hand over it again, and the improvement of the waveform was reproduced. He strangely discovered that a wall was needed to generate a stable Karman vortex. Therefore, he replaced more than 20 types of vortex generators of various shapes and dimensions in a pipe, and compared their waveforms every day.

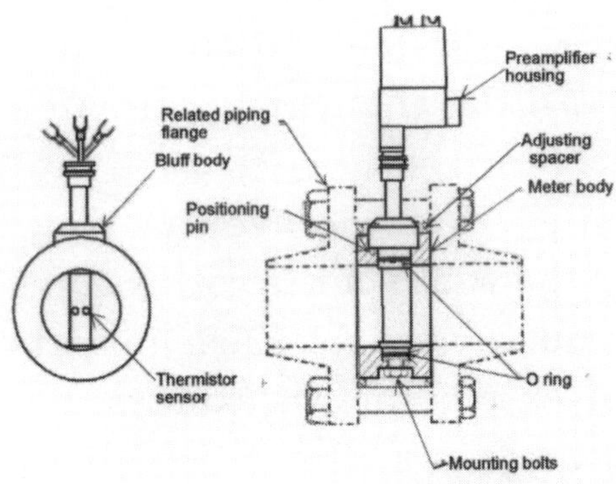

Fig.6.21 Thermistor type vortex flowmeter of Eastech Co. (35)

Then, he determined the best dimension of shape where the vortex was the most stable, and the thermistor type vortex flowmeter VS-21 shown in Fig. 6.21 was announced, and Eastech Co., was established.

Fig.6.22 shows the dimensional drawing of the vortex flowmeter manufactured by Eastech. Alan Rodely discovered that the vortex waveform is best in the triangular injection shape where the ratio of the inner diameter D of the meter

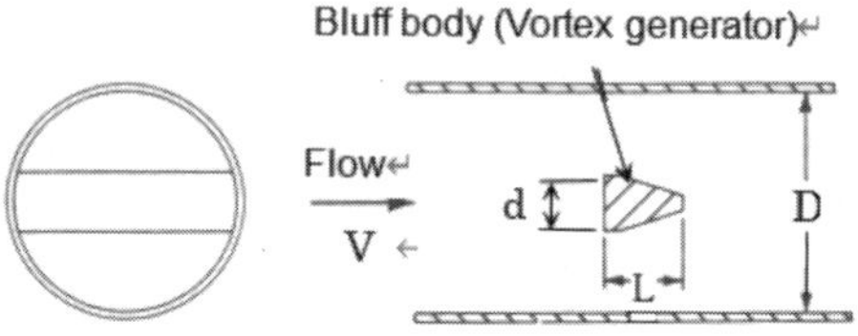

Fig. 6.22. Dimensional drawing of Vortex flowmeter (35)

body to the width d of bluff body d is d / D = 0.28. They found that it was stable and obtained a patent right. Here, the principle of the vortex flowmeter will be explained a little.

Assuming that the average flow velocity of the fluid is *V* and the width of the vortex generator is *d* in Fig. 6.22, the frequency *f* of the Karman vortex train formed behind the vortex generator is expressed by a simple equation as follows.

$$f = S_t \cdot \frac{V}{d} \text{-----------------------------------} (18)$$

Here, S_t is the constant known as the Strouhal number Fig 6.23 shows the outline of the electric circuit.

Fig.6.24 shows the results of calibrating this meter with air and water, which obtained good results with an accuracy within ± 2.0% of reading in the Reynolds number range of 360: 1.

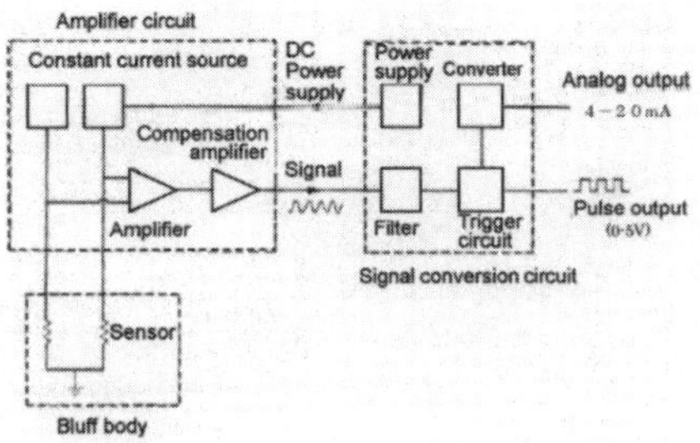

Fig.6.23 Overview of Electrical circuit

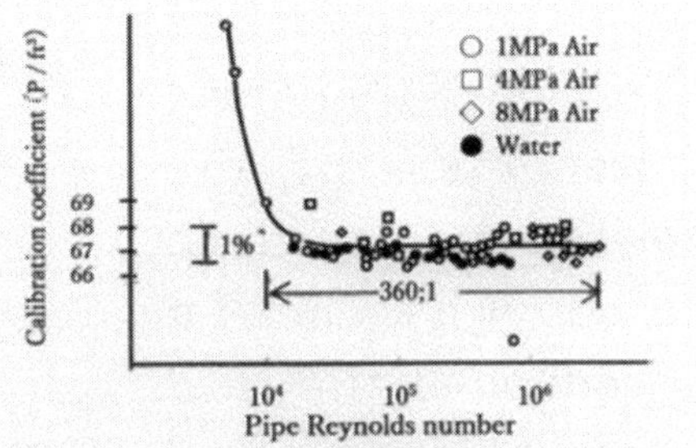

Fig.6.24 Characteristic of Reynold number of VS-21 (35)

In Japan, around 1968, Yokogawa Electric's Sahaji Tomoda, Hiroo Yamazaki and others developed a purge-type vortex flowmeter for flues. In this case, a cylinder with a diameter of 32 mm was inserted in the center of a large-diameter flue, where a vortex was detected using a platinum resistance wire, and the flow rate of exhaust gas was measured.

On the other hand, Eastech developed a movable piece detection method, a metho d of moving a sphere, plate, or cylinder up and down by vortex frequency vibration and detecting it with a pick-up coil for measuring steam flowrate and high temperature water. There were problems such as wear of the movable pieces.

Therefore, in 1977, Oval developed the strain gauge detector method built into a blaff body for vortex generator by Shoichi Hanzawa and others, and actively advertised that a digital orifice that made use of the pulse output was available.

As a result, Yokogawa would feel the crisis of the orifice meter, which was the main product of the flow department, and would work on the development of the vortex flow meter. In 1979, Itizo Ito and others developed the piezoelectric element type vortex flow meter shown in Fig. 6.25, which was commercialized.

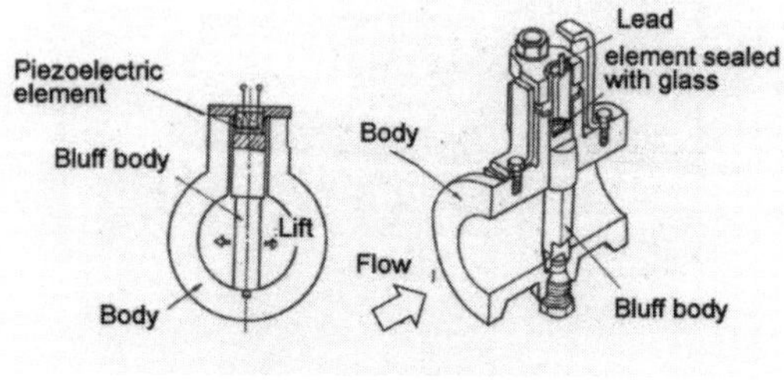

Fig. 6.25 Yokogawa Electric's Piezoelectric Vortex Flowmeter (YEWFLO) (36)

Its applicable temperature range was as wide as -40 to 300 ° C, there were no moving parts, and it was possible to measure steam, which would cause sensation in the world. Oval quickly fell into a predicament, because the strain gauge could not bear such a wide temperature range, but three years later Yokogawa, Naoki Matsubara et al. commercialized a piezoelectric vortex flowmeter as shown in Fig.6.26.

Because this piezoelectric element type vortex flowmeter was excellent in terms of performance and durability, Japanese companies took the initiative in this market and licensed it to Western companies, which would dominate the world market. In Japan, interest in vortex flowmeters increased, and JIS Z8766 "Flow rate measurement method

using vortex flowmeters" was established in 1989.
In the same year, the National Research Laboratory of Metrology (currently National Institute of Advanced Industrial Science and Technology) and several vortex flowmeter manufacturers, established the "Vortex Study Group.
This was because Eastech, Oval, Yokogawa Electric, Endress + Hauser, etc. use trapezoidal (or triangular) shapes for the vortex generators of vortex flow meters in the world. It was intended to determine a vortex generator with good performance and provide it to the world.

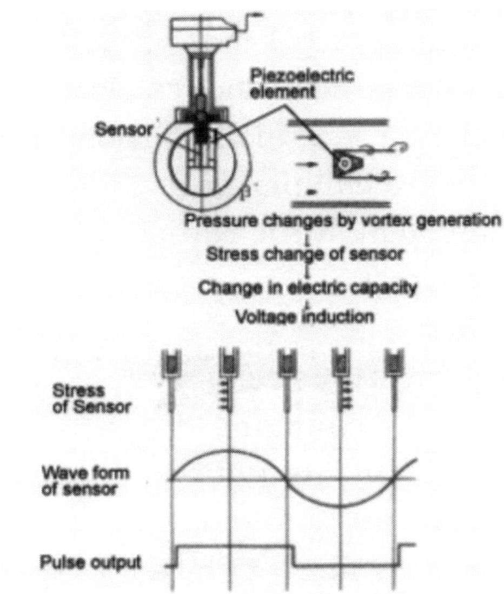

Vortex detection diagram by piezoelectric element

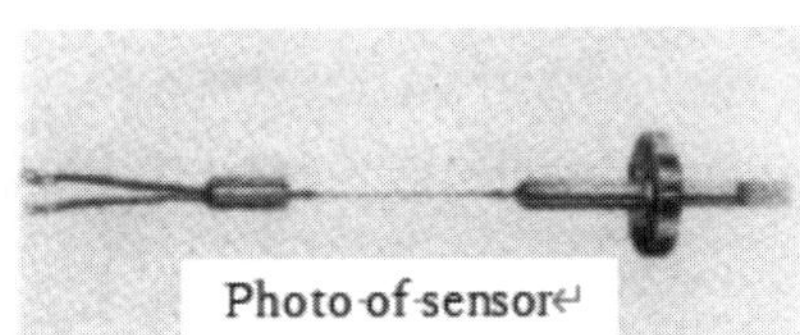

Fig.6.26 Oval Piezoelectric Vortex Flowmeter (Delta Meter) (37)

Masaki Takamoto and others announced the results at To

describe its outline, the geometric dimensions of the vortex generator of 150 mm vortex flowmeter is shown in Fig. 6.27, and the performance characteristics were examined by changing several dimensions of the flowmeter, using the standard equipment for large flowrate of liquid and air test equipment at the National Research Laboratory of Metrology.

At first, the characteristics were taken by changing d / D (W / D in this figure) in various ways, but it was found that it was the best in case of d / D = 0.28. Then, after that, it was fixed at d / D = 0.28, and the length (H) of the vortex generator was changed and the characteristics were examined. The results are shown in Fig. 6.28 (Note: These tests were conducted by facilities of AIST which will be described later.)

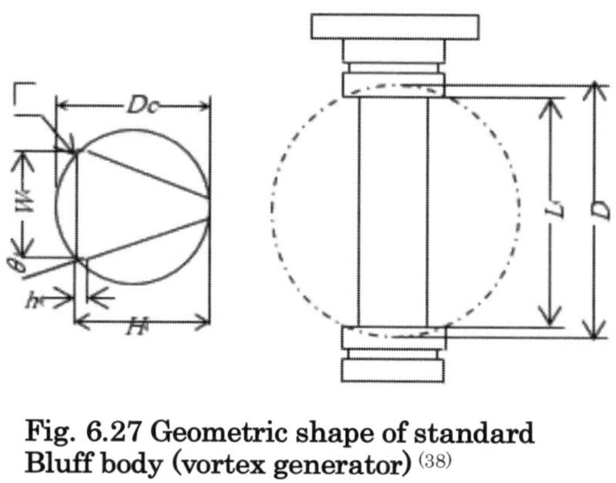

Fig. 6.27 Geometric shape of standard Bluff body (vortex generator) (38)

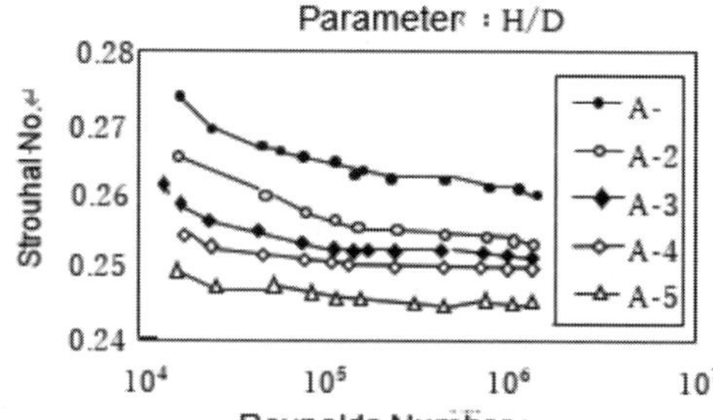

Fig. 6.28 Strouhal number characteristics of standard vortex flowmeter (38)

The Strouhal number S_t in Eq. (5) is the

function of Reynolds number, as a result of examinations, the linearities were best in case of A-4 and A-5. That is, it was announced that the linearity of the Strouhal number (constant) is good where the geometric shape is H / D = 0.35 to 0.40.

The geometric shape that Alan Rodely determined from only the waveform in the early experiments was d / D = 0.28, and the H / D was exactly the intermediate shape between A-4 and A-5.

What a wonderful pioneer's eye!

After that, capacitance type and ultrasonic type vortex flowmeters appeared. There are 14 companies currently affiliated with the Japan Measuring Equipment Industry Association handling vortex flowmeters.

Photo 6.24 Airflow sensor for automobile EGI air-fuel ratio control

As a unique application of the vortex flowmeter, in 1979, the oval ultrasonic type (air flow sensor) shown in Photo 6.24 was adopted to measure the amount of intake air for air-fuel ratio control of Mitsubishi Motors' electronic fuel injection system (EGI) for 10 years. After that, it was changed to a pressure sensor type vortex flowmeter. The number of vortex flowmeters delivered would be expected to reach millions, which will probably be the highest number of productions in the world as vortex flowmeters.

6.2.5 History of Thermal Flowmeters [29] [39]

The history of thermal flowmeters can be roughly divided into two groups. One is a method in which a heated object is exposed to a flow and the temperature is affected by the flow. There are a Hot-wire Anemometer for continuously heating and a Kata Thermometer for intermittent heating. The other evolved from the Thomas flowmeter.

As is well known, the Hot-wire Anemometer is used as a current meter, and thermocouples, thermistors, resistance wires, thin films, etc. are used as sensors.

The Kata thermometer is not well known until now, but it is used for environmental measurement of outdoor and indoor air flow.

In Japan, "Rakuten" that made a baseball team in the professional baseball Pacific League was also handling this type.

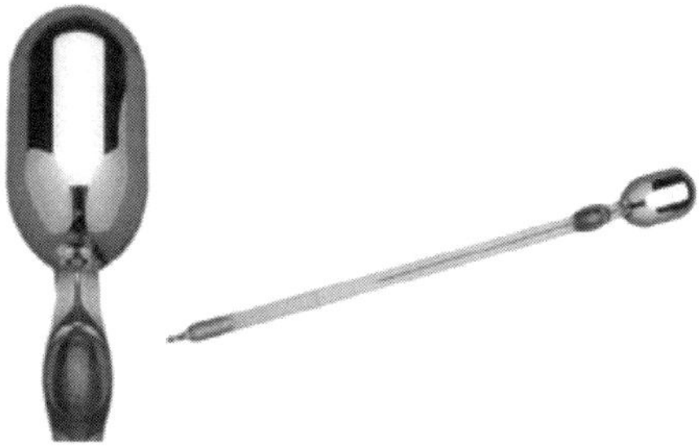

Photo 6.29 Current Kata thermometer

The principle is simple in that it a rod-shaped glass thermometer with the shape shown in Fig. 6.29, with alcohol enclosed in it.

This is heated to 38 ° C or higher, and the scales are added to 38 ° C and 35 ° C. It is a method that measures the time taken for the temperature to drop from 38 ° C to 35 ° C and then converts it into wind speed. In this first group of studies, a theoretical analysis was conducted by L.V. King of England around 1914.

The second group then connects to the thermal mass flow meters, which are now beloved for flow measurement in the semiconductor field. In the mid-1950s, NASA's space development was active in the United States. Because it was necessary to supply constant oxygen from the oxygen cylinder to the life support system,
which was indispensable for astronauts' spacewalks and extravehicular activities, regardless of the internal temperature and pressure, it is said that a bypass type thermal mass flow meter was born.

The first piece of literature about this was Raymond T. Doyle's "Metering Gas with a Heated Thermopile"(40) in 1957. According to this, it is based on the circuit of the US patent acquired by Hasting-Raydist Inc., to which he belonged, as shown in Fig 6.29 around 1951.

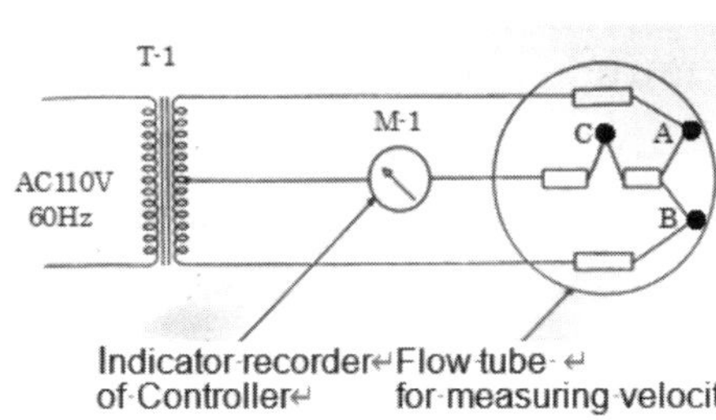

Fig. 6.29 Basic circuit of heating thermocouple flowmeter (40)

Sensors A and B form part of a bridge circuit with heated thermocouples, and C is a non-heated compensating sensor when temperature changes suddenly.

As shown in Fig. 6.30, the flowmeter has an orifice to divide flow and a flow tube installed in a bypass pipe. There was no mention that this thermocouple flow tube was used in the life support system of NASA's astronaut, but it might have been adopted because of the timing agreement. It might not have been announced because of military secrets.

Fig. 6.31 was invented by J. H. Laub in 1947, where two Heat Sinks were used to heat a tube wall.

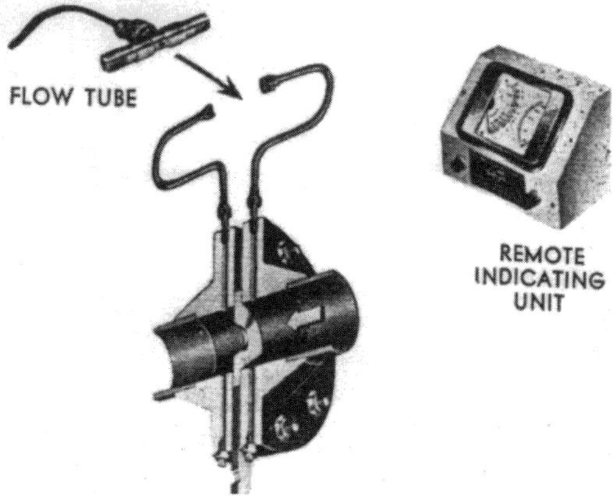

Fig. 6.30 Installation of thermocouple flow tube [40]

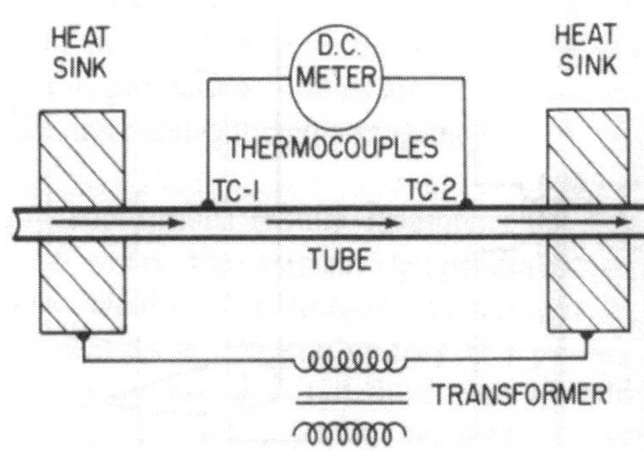

Fig. 6.31 Main pipe heated type [39]

Fig. 6.32 is a structure announced by J.M. Benson et al. in 1970 similar to a current bypass tube type thermal mass flow meter.

It is said that a thermal mass flow meter was first commercialized by John Olin and Jerry Kruz of the United States. (41) They were classmates at Stanford University, and joined TSI (Thermo System Inc.) in 1968 after graduating. The two were concerned that the Hot-wire Anemometry was very fragile for industrial use, so the first product they developed was a portable anemometer and its transmitter.

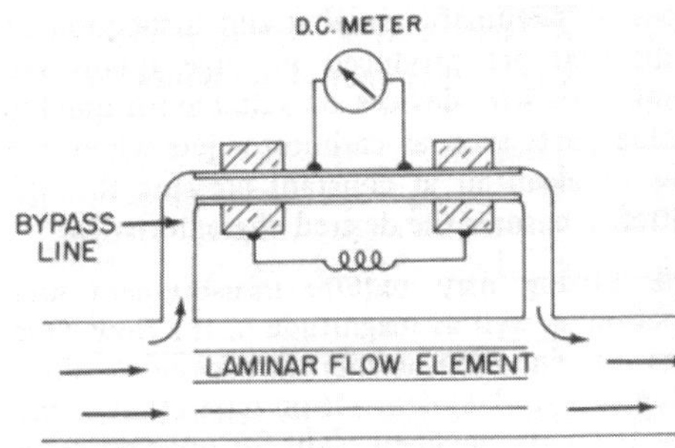

Fig. 6.32 Bypass heated type (39)

Then in 1975, they left TSI and developed an air sampling product that measured particles in the air and their distribution, laying the foundation for Sierra Instrument. In 1977, Olin ran Sierra, while Kurz founded Kurz Instrument. In the 1980s, Olin systematized a Bypass Type Thermal Flowmeter, and Kurtz systematized a DC Type Thermal Flowmeter. The bypass flow type is now the mainstream of thermal flowmeters.

Therefore, the principle of the Bypass Thermal Flowmeter will be briefly explained.

Assuming that volumetric flow rate flowing through a capillary is Q_b, the gas density is ρ, and the constant pressure specific heat is C_P, the temperature difference ΔT (output signal V) between two temperature sensors before and after, the heating element is expressed by Equation (16).

$$V = \Delta T = K \cdot C_P \cdot \rho \cdot Q_b = K \cdot C_P \cdot Q_m \quad (16)$$

Here, Q_m is a mass flow rate, which is why it is called a thermal mass flow meter.

Since constant pressure specific heat is different for each gas, the output signal is different for the same mass flow rate. Recently, flow sensors made of silicon wafers and thermal flow meters for liquids have also been commercialized. Nowadays, there are as many as 9 companies handling thermal flowmeters in Japan.

6.2.6. History of Coriolis Flowmeter (42)

The Coriolis flowmeter is the newest principle for industry. This flowmeter is based on Coriolis force principle found by French mechanical engineer Coriolis (Gaspard Gustave de Coriolis, 1792–1843), as described in

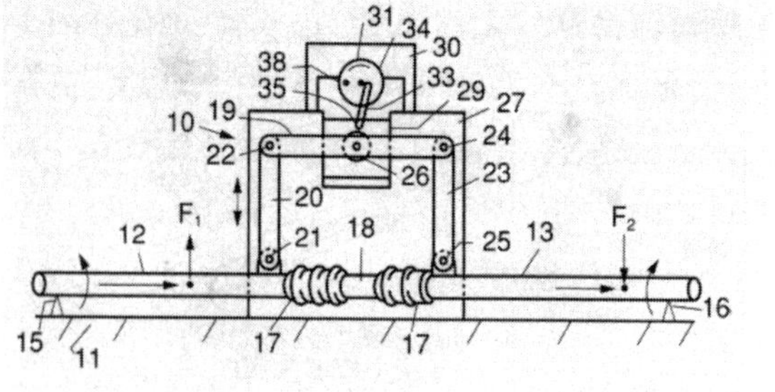

Fig.6.33 Smith's first patent for Coriolis Flowmeter in 1975

Section 3.1.2.(2). The first pioneering technology for this flow meter in the world
was the Sipin's patent in 1964 (US Patent No. 3,355,944).

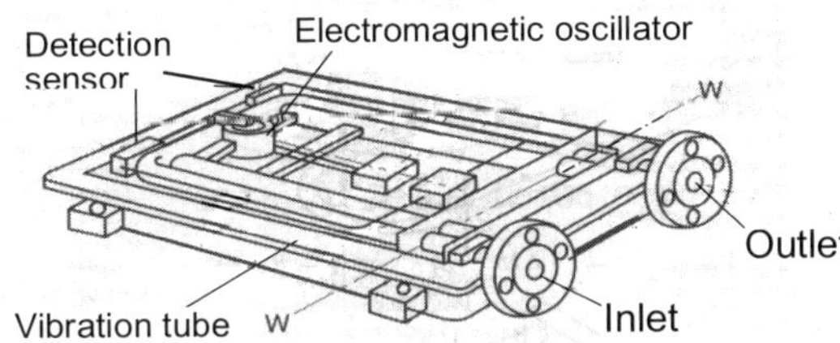

Fig.6.34 Initial Micro Motion Flowmeter

After that, J.E. Smith of the United States applied for patents concerning this type flowmeters one after another. Fig.6.33 is the Smith's first patent for Coriolis Flowmeter where a flow tube is mechanically vibrated in 1976.

He established Micro Motion Co., Ltd, and started marketing direct mass flow meters around 1977.
An initial one is shown in Fig. 6.34. This was called the C type at that time, and there was only one vibrating tube.
When the vibrating tube is vibrated by an electromagnetic oscillator, the tube bends and vibrates at

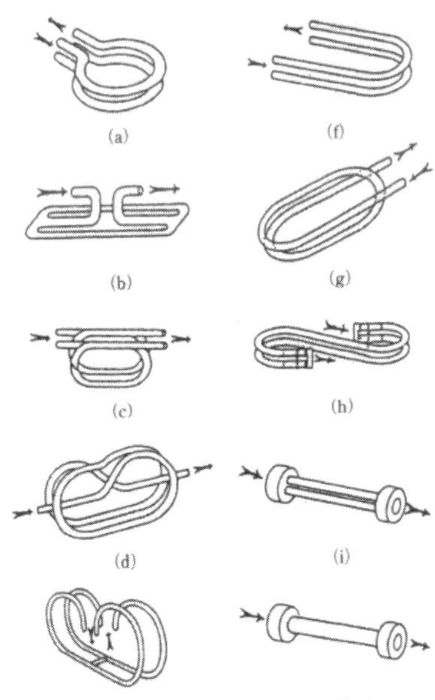

Fig. 6.35 Typical examples of tube shape (40)

its natural frequency along the W-W line. When a fluid flows through this tube, a vibrating tube causes not only bending vibration but also torsional vibration according to the Coriolis force. The torsion angle of this torsional vibration is detected by two sensors, and the principle of the Coriolis flowmeter is that this torsional angle is proportional to the mass flowrate of fluid. The detection sensor initially used two photo sensors, but later began to use two electromagnetic pick-up coils like this figure.

After the advent of Micro Motion flow meters, various tube-shaped Coriolis flow meters as shown in Fig. 6.35 appeared. in order to avoid Smith's patent

Around 1982, Micro Motion Co., Ltd. introduced what was called a D-shape with two parallel U-tubes (f), and its performance would be further improved. This factor is due to the stabilization of vibration by resonance phenomenon as well as tuning fork. Its detailed figure is shown in Fig.6.36. (b), (c), (d), (e), and (h) in Fig.6.35 were devised to escape the patent of this parallel tube type Coriolis flowmeter (f). In 1987, the Coriolis flowmeter with two parallel straight tubes shown in (i) was announced by Flowtech, a subsidiary of Endress + Hauser, and caused a sensation.

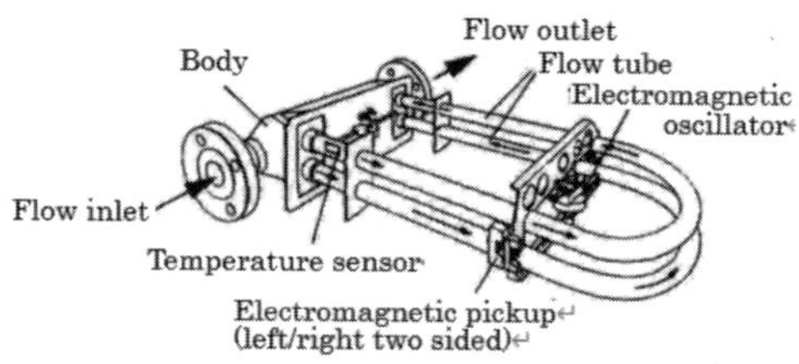

Fig. 6.36 Two parallel tube types Coriolis flowmeter [42]

Inspired by this, Krone and Oval each other announced (j) single-tube Coriolis flowmeter one after another after 1992. Figure 6.37 shows the Oval one. In principle, this flowmeter can measure any fluid as long as stable torsional vibration is generated by the Coriolis force, and it can also measure slurry and gas. Since there are no elemental parts in the tube and the fluid density can be measured from the natural frequency, it is attracting attention as a composite instrument, and ISO standards and JIS standards had already been issued. Here, the measurement principle is explained based on Fig. 6.38. As shown in the figure, when a fluid is flowing from the inlet to the outlet of one tube at a flow velocity *V*, if this tube is vibrated at a circular frequency ω, the tube receives the next Coriolis force F_C.

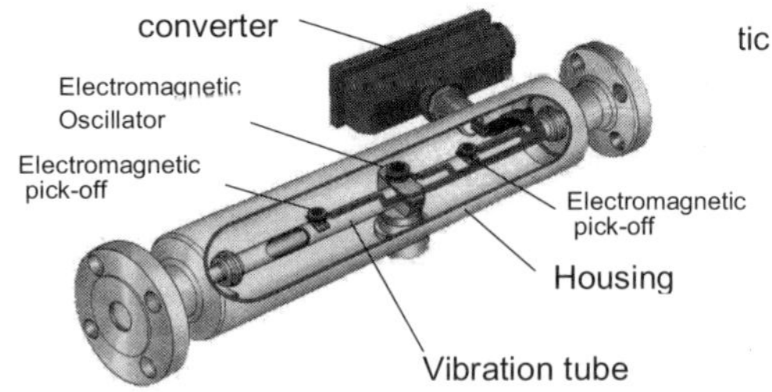

Fig. 6.37 One -straight tube type Coriolis flowmeter of Oval (43)

$$F_C = 2 \cdot \Delta m \cdot \Omega \cdot V \quad \text{------}(19)$$

Here, Δ *m* is the unit mass of the fluid. If following formulas in middle are omitted, finally the mass flow rate Q_m can be expressed as

$$Q_m = \frac{k_S \cdot \tau(1-\omega^2/\omega_S{}^2)}{2K \cdot d^2} \quad \text{------}(20)$$

Here, the meanings of the symbols are as follows.

Ω: The natural Bending circular frequency

ω_s: The natural frequency of twisting.

k_s: The spring constant of the tube.

τ: Time difference detected by the detection sensors on both sides.

K: Constant.

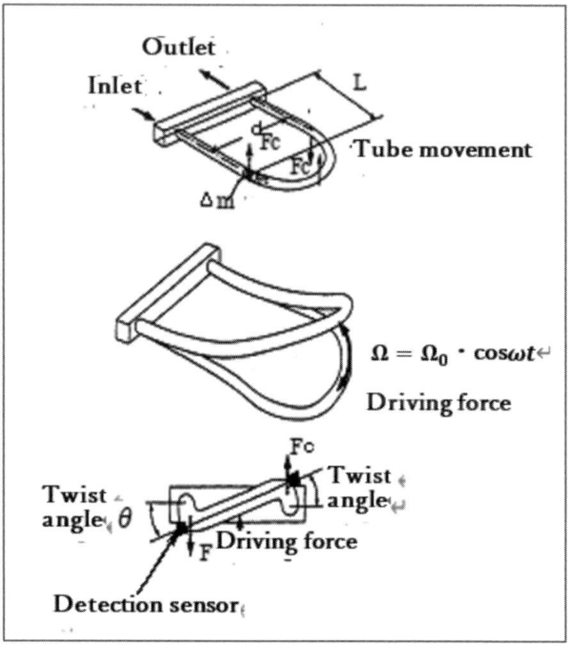

Fig. 6.38 Operating Principle of Coriolis Flowmeter [(41)]

types are (i), (j), and the bending ty In the Fig.6.36, the straight tube pes are (a), (b), (c), (d),(e),(f),(h).

The loop shape means a shape twisted like (g).

Nowadays, more than 8 companies currently handle Coriolis flowmeters in Japan.

6.2.7. Postwar History of Turbine Flowmeters

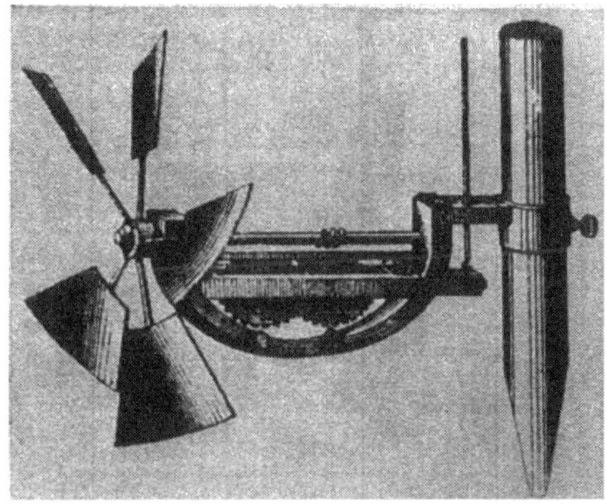
Fig. 6.39 Prototype of Waltman meter (44)

The history of the turbine flowmeter is described in section 5.6.2, but the root of turbine flowmeters goes back to the Waltman Meter in 1790. Since this prototype was obtained, it is shown in Fig. 6.39(44). This was depicted in a German handbook "Die Hydraulik" published in 1878, and as can be seen from the figure, the flow could be measured only in rivers or open culverts. The rotation of turbine was transmitted to the integrator by gear trains.
In addition, it is said that calibration had been already performed for accurate flow measurement.

Now, the postwar trend is described. Some turbine meters were used to measure the flowrate of fuel oil for US Air Force rocket engines in the 1960s, but there were problems that the integrated values on site and the calibration results at laboratory did not match. Therefore, ICRPG (US Chemical Rocket Propulsion Group) developed an analysis program for the development of turbine flowmeters that took into account parameters that might affect the performance of the meters.

In April 1967, the US Army Missile Command in Alabama also made a deposit with Greyrad Co., Ltd, in New Jersey, for research on turbine flowmeters. The results of this study were reported to HQ six months later [45]. This was 200 pages long and probably the most systematic and analytical technical document of turbine flowmeters in existing literature. Companies such as Fisher & Porter, Potter, Cox, and Foxborough were collaborating on this study. [46]

As one of the achievements of this project, API standard 2534 was published in 1970, and turbine flowmeters became popular in the oil industry.

The shape of rotor blades for turbine flowmeters are roughly divided into two types: straight blades derived from Potter and spiral blades derived from Fisher & Porter and Rockwell. Fig.6.40 shows the structure of Tokico's Potter meter.

As already mentioned in section 5.6.2 (1), in this meter has a hydric balance made by the shape of upper and lower cones. The thrust force from the inflow direction is reduced to improve the performance of the flowmeter.

It seems that the principle of the turbine flowmeter is obvious, but it will now be explained. When the fluid flows into a housing, it hits the rotor blades and the rotor rotates. The rotation is detected by an electromagnetic pick-up coil. Outputs are as many pulses as there are blades in one rotation.

If the number of pulses is N, the volume Q that has passed through the meter is expressed as

$$Q = K \cdot N \text{------------} (20)$$

Here, K is called the pulse constant (L / P). Photo 6,25 is an example of a modern oil lorry shipping site with turbine flowmeters designed and constructed by Oval. 13 companies currently handle turbine flowmeters in Japan. 12 companies handled normal axial types and 10 companies' tangential types. Normally the former types are used for industrial turbine flowmeters, and the latter ones are used for domestic water meters. The description so far is intended to describe the history of various flowmeters, but

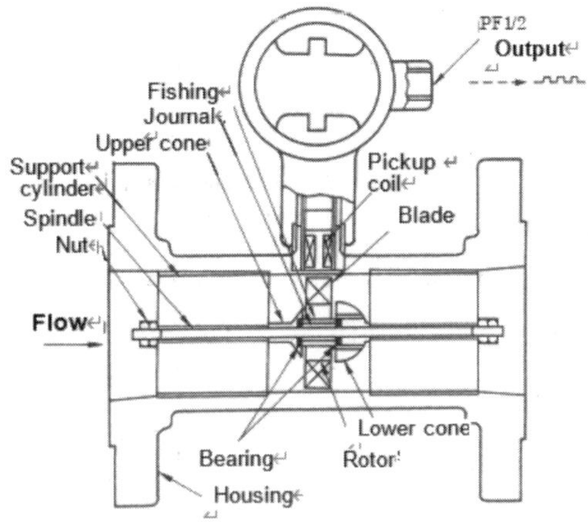

Fig. 6.40 Structure of current turbine flowmeter (46)

Photo 6.25 Lorry shipping system with turbine flowmeter (47)

I think it does not cover all of them. Since I am writing this mainly as a pioneer of various flowmeters, I am sorry some new technology may be missing.

6.2.8. History of Process Automation (29) (49)

Now, industrial flowmeters are often installed in plants, and there are aspects that cannot be talked about without instrumentation or process automation.

Therefore, the history of instrumentation after the war is now reviewed a bit.

Automatic control, which measures and automatically controls industrial quantities such as temperature, pressure, and flowrate, began in the 1920s American oil refining process. In the 1930s, it was a local system that simply installed a large mechanical controller (for example, manual valves) on site.

In the 1940s, the development of large pneumatic instruments began in the United States, and in the 1950s after the end of the Second World War, pneumatic PID controllers were introduced, and the signals of pneumatic industrial instruments were unified to 3 to 15 psi (20 kPa to 103 kPa).

In 1955, small pneumatic instruments were completed and mass-produced. However, in 1958, small electronic instruments appeared and were put into practical use. Here, it became controversial whether instrumentation is pneumatic or electronic. Initially, it was recognized that pneumatic instruments were more reliable than electronic ones, but what prompted the shift to computerization was the expectation that something would be more accurate and the premonition of the arrival of process computers.

Computers began to be used for process control in the 1960s, and in 1967, DDC (Direct Digital Control) was developed in Japan, but it was an expensive system and did not spread.

In 1971, the transmission signal, which had been a concern for many years, was unified to DC4 to 20mA.

In 1975, distributed control systems (DCS) appeared in the United States and Japan almost simultaneously. In Japan, Yokogawa announced CENTUM and Honeywell (Yamatake) announced TDCS2000.

Hokushin Electric Works has been working on DDC in the past, but in 1976 it announced the 900 / TX (See Photo 6.26)

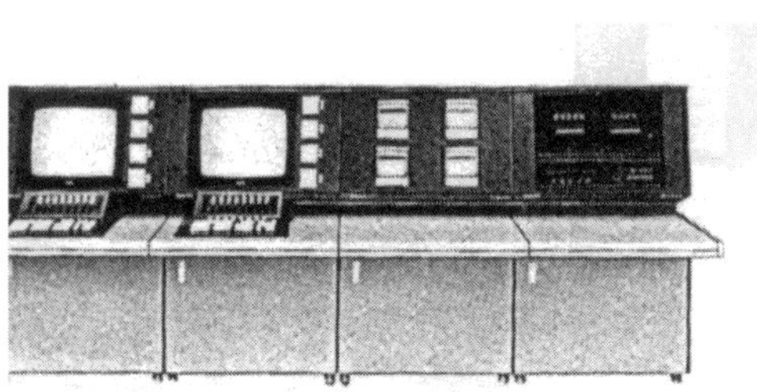

Photo 6.26 Hokushin Electric Works Limited Control System 900

and followed the movements of both companies.

The success of DCS was due to its low cost and high-performance microprocessor.

In September 2001, at the ISA show, Emerson's Vice President John Bella boldly stated that the size of the automation market was 2.8 trillion yen of which broke down as 14% for Emerson, 11.6% for ABB, 8.9% for Siemens, 8.4% for Invensys, 7.6% for Yokogawa, and 7.5% for Honeywell. It seems that there was considerable counterargument to this ranking depending on the viewpoint$_{2010}$Honeywell was acquired by Allied Signal in 1999, but barely retains its name. The transition of market size for process automation in the past can be represented by Fig. 6.41.

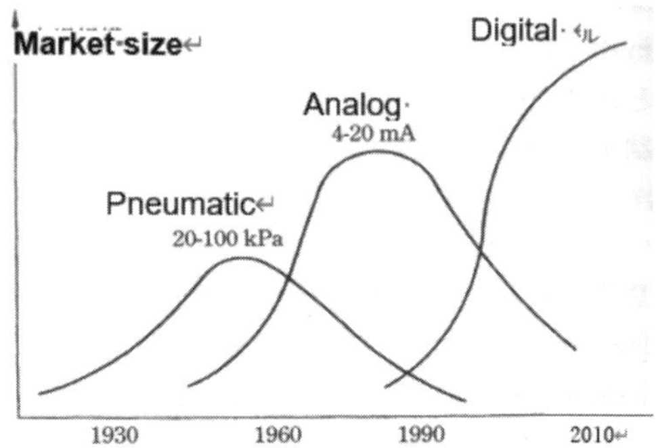

Fig.6.41 Transition of transmission Signal

(50) Yoshio Tsukada "Globalization and its Memories" Measurement Technology Vol.31. No.5, 2003.

6.3 History of World and Japanese Flow Standards

6.3.1 History of the Standard Tank

The history of flow measurements has been described for a long time, but the one of flow standards has not yet been systematically described. There is no doubt that contai(hé):s such as measures and flasks were used to check the accuracy of flowmeters when they first appeared.

In China, it is said that volumetric measures existed already in the era of Zhou Dynasty (1122 to 250 BC). Qin Shi Huang unified the weights and measures in 221 BC, but the standard container is not clear. Liu Xin made a standard container as shown in Fig. 6.42 at the time of Wang Mang of Han in 1 to 5 AD. It was a clever one that had five types of standard instruments: Yuè, hé, shēng, dòu, hú. If the volume needs to be measured by using hú or Yuè, the container is turned over and it can be measured. (50)

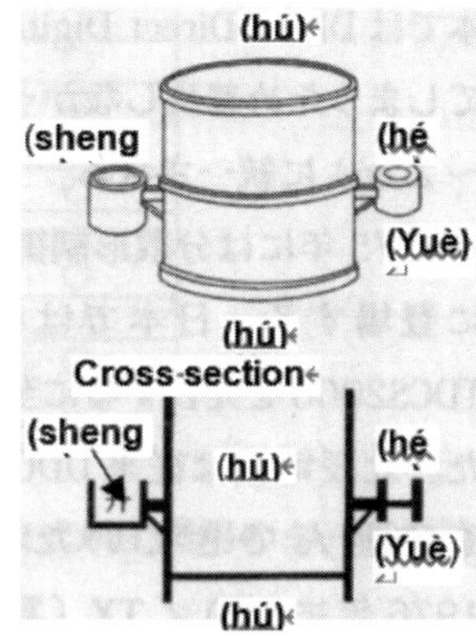

Fig.6.42 Standard measure made by Liu Xin (50)

Photo 2.13 in section 2.2.3 also shows the hydrometer of the Roman Accademia del Cimento around 1651 as a historical fact.

Furthermore, in 1888, Herschel used a standard tank to calibrate the Venturi meter in Fig.4.26 in Section 4.2.4, and in Fig.4.16 in Section 4.2.1, several types of water meters were calibrated in the 1875s, probably by using standard containers.

In the 17th, Fig. 5.52 in Section 5.5.3 shows the calibration device using the reference tank used by the ASME Flowmeter Expert Committee in 1940 for the volumetric flowmeter test. These can be said to be a series of flows concerning flowrate standard from ancient times to prewar days.

6.3.2 History of Pipe Prober (Reference Volume Tube)

After the war, the flowrate of oil handled at oil refineries became extremely high at 500 to 1,300 m^3 / h, so how to

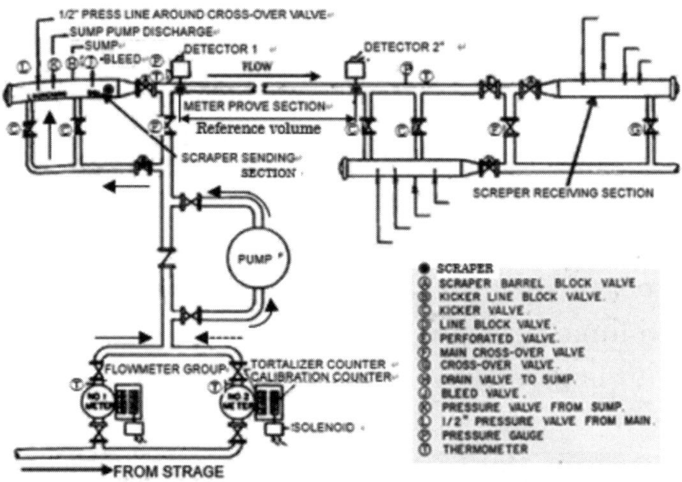

Fig. 6.43 Proving device of US Shall Oil Co., "Miracle Miles" (51)

calibrate the flowmeter became a problem. It was very difficult to calibrate with the standard tanks so far.

Fig. 6.43 shows a probing device called "Miracle Mile" from the US Shell Oil Co., announced in 1954, which was the first one of the recent pipe probers.

To explain a little about this device, flowmeter groups at the bottom are calibrated using the reference volume of proving device. A scraper is sent out from the scraper sending section by operating valves. The scraper is a disk or cup shape of 76mm diameter made from a little soft rubber that is slightly larger than the inner diameter of pipe in meter prover section so that it becomes tighter.

The scraper is put on standby in the front of detector 1 as shown in Fig.6.44. Next, a kicker valve © is operated to flow calibrating liquid coming from the storage tank through a flowmeter (No.1 or No.2).

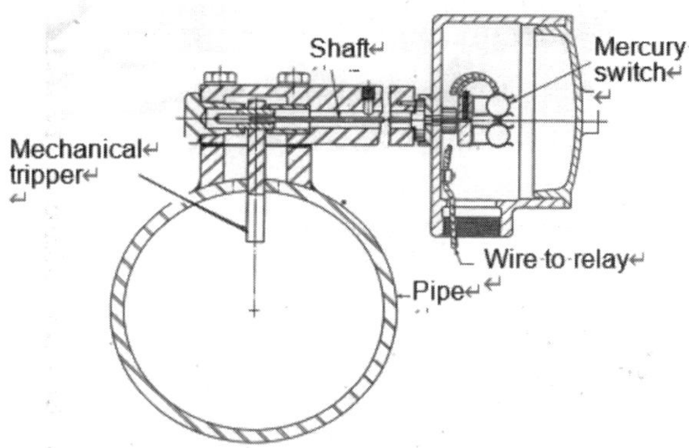

Fig. 6.44 Detector structure

The scraper switches on detector 1 and it moves by flow downstream, then switches off detector 2 and fits in the scraper receiving unit. Pipe volume between detector 1 and detector 2 becomes the reference volume, and the indicated values of the counter of the flowmeter linked to the switch on and off may be compared. As per this figure, the scraper can be returned to the sending section by operating kicker valves ©.

Originally in the United States, based on the results of multipurpose utilization technology of oil transportation pipelines and cleaning technology inside pipes, a probing system like the "Miracle Mile" were completed, and remodeling was repeated for 9 years. Later, in 1963, the API standard 2531(53) pipe prober was established.

On the other hand, in Japan as well, in June 1969, a part of the standard inspection ordinance and standard inspection rules based on the Measurement Law was amended, and the pipe prober was allowed to be used as a standard. Fig.6.45 shows the bidirectional pipe prober among the various probers of API 2531.

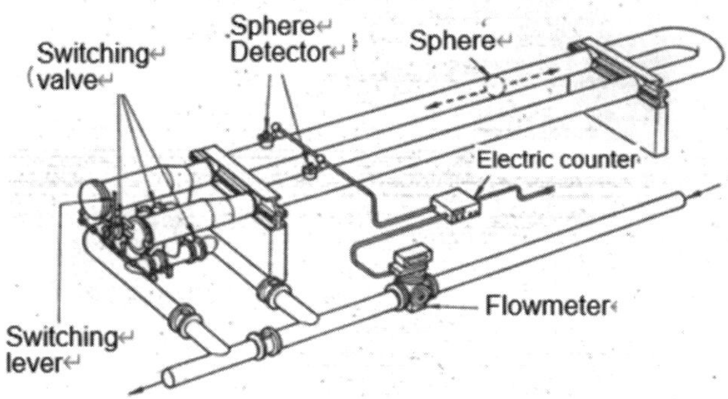

Fig. 6.45 Bidirectional pipe prover (53)

This compensated for the shortcomings of the "Miracle Mile", making the moving object as a sphere, that is, as a ball, allowing the meter to be calibrated by moving it in both directions as many times as necessary without removing it. The reference volume is the pipe volume between the sphere detectors.

6.3.3 Japanese Weighing Type Oil Flow Standard Facility (65)

In Japan, the National Institute of Advanced Industrial Science and Technology in July 2002 opened a standard oil flowrate facility shown in Photo 6.27. According to the recent Homepage of Liquid Flow Standard Group of AIST, the oil flow calibration facilities for kerosene, light oil and spindle oil cover the range between 0.02 L/h and 300 m^3/h with uncertainties between 0.020 % and 0.078 %.

Photo 6.27 Japanese weighing oil flow standard (65)

These facilities are roughly divided into for large, medium and small one.

A large facility is shown as Photo 6.27 and its imaginary figure below is compared to the size of man.

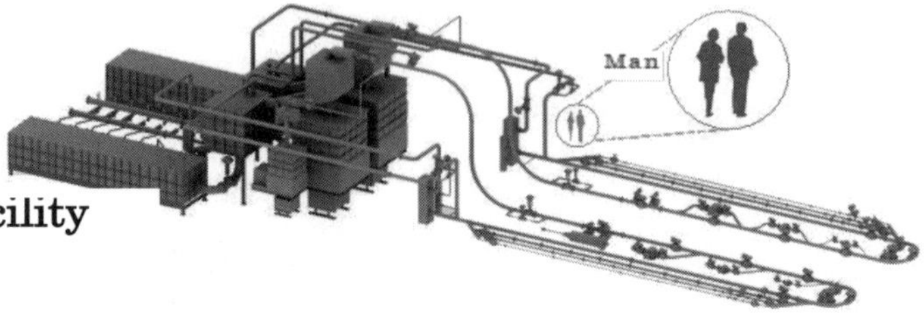

● **Large facility**

General future

Flowrate	3 to 300 m^3/h
Uncertainty	0.020 % (k=2): for mass flow. 0.030 % (k=2): for volumetric flow
Working Liquid	Kerosene, Light Oil
Calibration Method	Static and Gravimetric Method with Flying Start and Stop Method
Temperature	15 to 35 °C, Fluctuations for 10 min.: 0.05 °C or less
Pressure	200 to 600 kPa
Test Line	Pipe Diameter: 50 to 150 mm, Straight Pipe Length: 100 D or More
Main Equipment	Weighing Scale (10 t, 1 t) Diverter (10 t, 1t) Density Meters
Check meter	Servo PD Meters (SFL)

● **Medium facility 2**

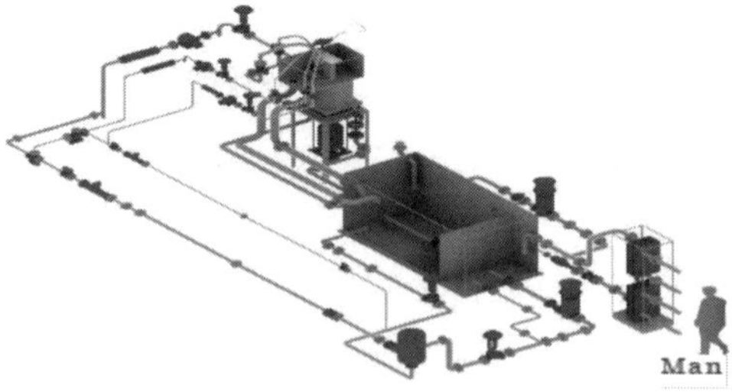

General future

Flowrate	0.1 to 15m³/h
Uncertainty	0.020 % (*k*=2): for mass flow 0.030 % (*k*=2): for volumetric flow
Working Liquid	Kerosene, Light Oil, Spindle Oil (Exchange), Industrial gasoline
Calibration Method	Static and Gravimetric Method with Flying Start and Stop Method
Temperature	10 to 35 °C, Fluctuations for 10min.: 0.05 °C or Less
Pressure	200 to 700 kPa
Test Line	Pipe Diameter 4 mm to 40 mm, Straight Pipe Length 80 D or More
Equipment	•Weighing Scale : 32 kg x 1 •Rotary Double Wing Diverter •Density Meters •Check Meter: PD Meter

● Small facility

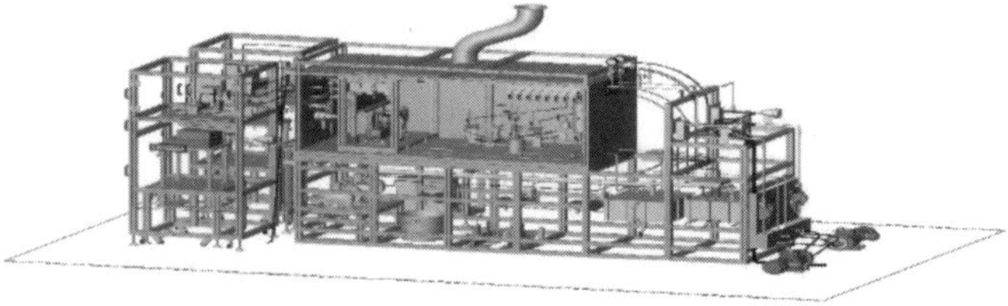

General future

Flowrate	20 mL/h to 100 L/h
Uncertainty	0.020 to 0.050 % (*k*=2): for mass flow 0.064 to 0.078 % (*k*=2): for volumetric flow
Working Liquid	Kerosene, Light Oil, Industrial gasoline
Calibration Method	Static and Gravimetric Method with Flying Start and Stop Method
Temperature	15 to 35 °C, Fluctuations for 1 hr.: 0.05 °C or Less
Pressure	up to 400 kPa
Test Line	Pipe Diameter 1/4 inch to 1/2 inch

Main Equipment	Gravimetric system with diverter •2 separate systems for different working liquids •calibration flow range 1 L/h to 100 L/h • 2 kg weighing scale (2 units) •conical rotating double-wing divert

From this data, the greatness of oil standard facilities in Japan today can been seen.

Let's look back flowmeter tested data or equipment in this book.

In the data in Fig. 4.16 from around 1876 in Section4.2.1, there is no description of the testing method, but it would be a comparison with the water scale (amount of water in the water tank). Next is the durability test result of Fig.4.22 from around 1893 in section 4.2.2 that would be also tested by the water scale.

The test equipment was first described in Fig.4.26 in section 4.2.4, when Herschel tested the Venturi meter in water in 1888. This was also due to the comparison with the water scale. Next, Fig. 5.14 in section 5.2.1 was an important experiment that the coefficient of variation (Z) of gas (natural gas) was different from ideal gas that was determined using a weight-type device, and this was in a report to ASME by Robert F. Earhart in 1914.

Finally, as described in section 5.5.3, as shown in Fig.5.52, the United States introduced a wonderful systematic modern oil testing facility.

So, next, I would like to investigate the historical test equipment for air flowrate.

6.3.4 History of Gas Flow Standard

Compared to the liquid flowrate standard, the gas flowrate standard has a shorter history. As mentioned in Section 5.2, the accuracy of natural gas flow measurement was discussed by Weymouth in 1914.

The first method mentioned was what led to the Constant Volume Tank in the current volume standard.

In 1950 C.M. Halsall et al. discussed "Design of a Pvover for

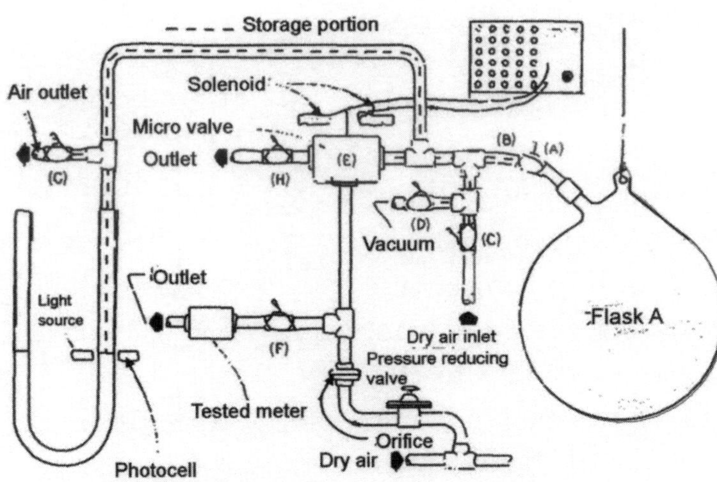

Fig.6.46 Gas test equipment of Evans (67)

Mass Measurement of Gas" (66). This was a method of

measuring the weight loss before and after the air flow from a high-pressure gas container.

In 1965, Evans applied this method to small gas flow measurement and announced the method of measuring gas mass flowrate within 0.1%, as shown in Fig. 6.46(67). However, since there are some points which are difficult to understand, it will not be described in detail.

Thereafter, modern gas flow and flow velocity standards will be described based on Takamoto's commentary. (68)

The Dutch flow standard is famous for having a device that can calibrate up to 6.1 MPa and 2.5 million Nm^3 / h by using natural gas, and a bell prober shown in Fig. 6.47 was used as the first flowrate standard.

The principle is as follows.

Since the internal volume of a bell (sinking bell) is sealed in an oil tank, the flow of gas comes in and goes out according to the speed of the bell moving up and down. This flow is used as the standard for volumetric flow.

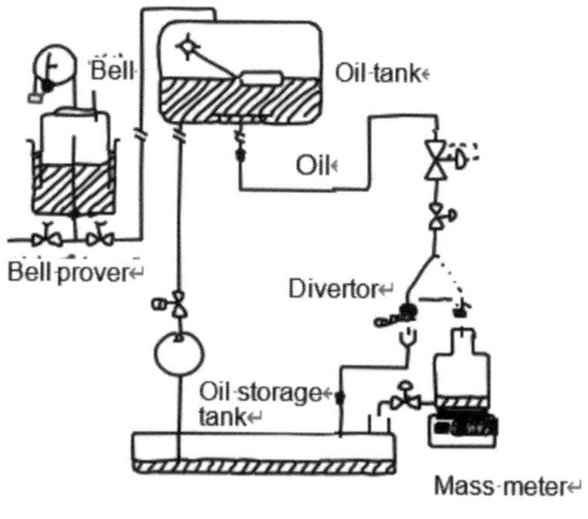

Fig.6.47 Flowrate standard by bell prober (68)

The internal volume of the bell is replaced by the volume of oil, which is determined by measuring its density and mass.

Today's Bell Prover can be measured under atmospheric pressure with a flow range of 1 to 100. (1993)

As shown in Fig. 6.48, the American flow calibration organization CEESI pressurizes gas in a pressure vessel made of titanium alloy and measures its mass with a balance.

This pressure vessel is separated and connected to a test line, the line is running-in with another pressure source, and then the valve is switched to allow gas to flow from the pressure vessel to test pipe line for a certain period of time. After the test is completed, the pressure vessel is separated from the line again, its weight is measured with the balance, and the mass flowrate is calculated from the difference from the initial value. (1993)

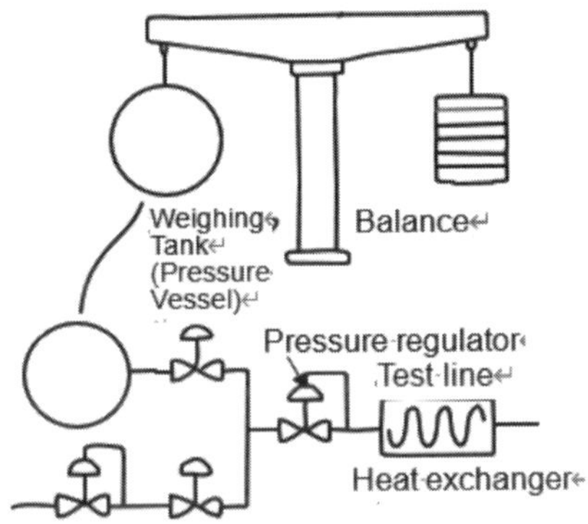

Fig.6.48 Flow rate standard by weighing tank (68)

In the UK and Norway, automation is being attempted with a mass meter that uses a gyro instead of a balance.

6.3.5 Introduction of AIST (Japan)

The National Institute of Advanced Industrial Science and Technology (AIST), one of the largest public research organizations in Japan, focuses on the creation and practical realization of technologies useful to Japanese industry and society, and on "bridging" the gap between innovative technological needs and commercialization.

Now, the latest fluid (air, water, and petroleum) flow standard facilities, within this organization will be described, "The National Research Laboratory of Metrology (currently AIST) announced a method using a constant volume tank as a gas flowrate standard in 1993. In this method, the volume of constant volume tank is previously measured by mass measurement method using high-purity nitrogen gas. As shown in Fig. 6.49, a constant volume tank is circulating

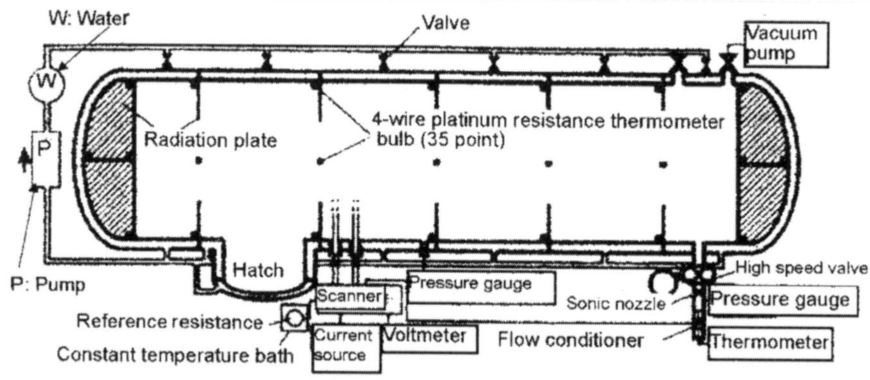

Fig. 6.49 Gas flowrate standard by constant volume tank (68)

water in the gaps between the double walls to make temperature distribution as uniform as possible. In addition, lots of platinum resistance thermometer bulbs are used to measure the average temperature inside the tank." (68)

In the field of medium flowrate in gas, specified standard have been realized by a method using a constant volume tank with the volume of 13.4m^3 as shown in Fig. 6.50.
The system by the constant volume tank is used to calibrate sonic nozzles. That is, after the constant volume tank is depressurized, air is sucked through a sonic nozzle for a certain period of time.

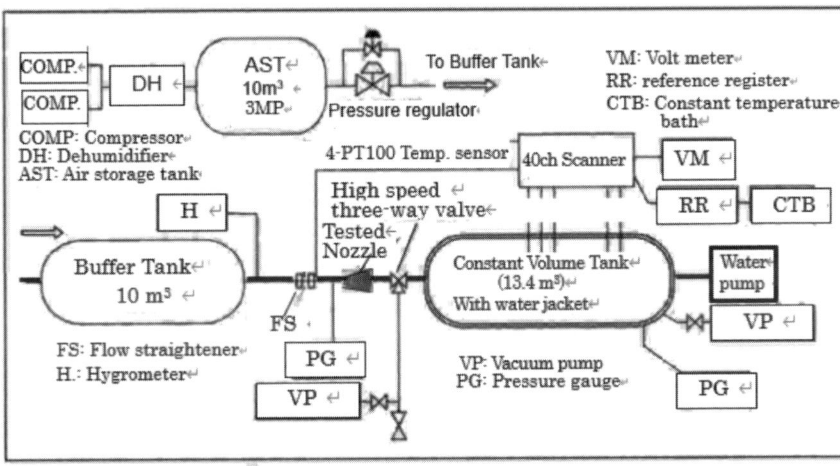

Fig.6.50 Calibration of sonic nozzle by constant volume tank (29) (69)

Then, the mass of intake air is calculated from the temperature and pressure before and after suction and the stated equation of the air. Finally, by calculating the mass divided by suction time, a volumetric flowrate passing through a sonic nozzle can be obtained.

Since the flowrate of the sonic nozzle has a feature not affected by the state of the downstream side, it is possible to keep the flowrate constant by utilizing this characteristic.

At present, sonic nozzles were used in other countries as transfer standards with the least uncertainty in a such flowrate range.

The pressure range of sonic nozzles that can be calibrated by this constant volume tank system is 100 to 500 kPa, and the flowrate range is 5 to 200 m^3 / h.

Figure 6.51 shows the closed-loop test equipment for calibrating each flowmeter, using the sonic nozzle calibrated

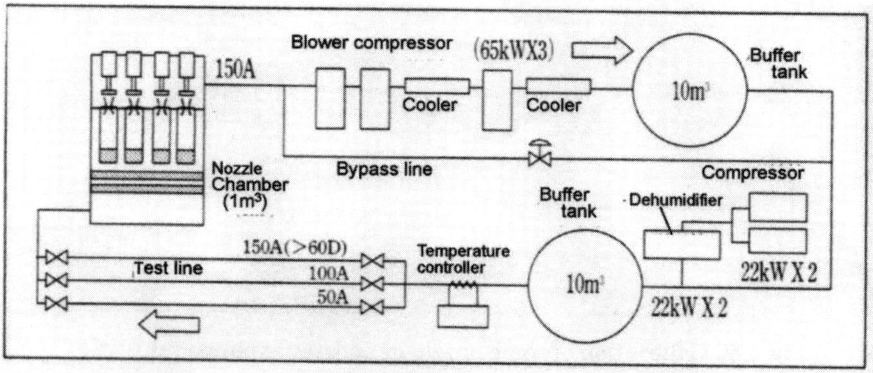

Fig. 6.51 Closed loop flow meter calibration equipment [29]

by the above-mentioned constant volume tank system as the secondary standard.

Dry air is used as the fluid, the maximum pipeline pressure is 0.5Mpa, and the maximum flowrate is 1,000m^3 / h in actual flow rate.

Flow is driven by three blowers and the pressure drop of the whole closed loop is allowed up to 300 kPa. To reduce the uncertainty of temperature measurement, a large capacity heat exchanger is installed in the pipeline, and all the equipment is placed in an air-conditioned room.

6.3.6 Japanese Calibration Equipment of Gas Flow Standard Research Group

According to the recent homepage of the Gas Flow Rate Standard Research Group of AIST, this group works on standard setting and supply of gas flowrate and gas velocity, and is developing the measurement technology of pulsating gas flowrate.

Each standard has different gas flowrates other than air at small flowrate. Medium flowrate is limited to air one only. The gas flowrate can be divided into 0.005 to 200 g / min at the small flowrate and 5 m^3/h to 1000 m^3/h at medium flowrate. The calibration system for the small gas flowrate shows following Fig.6.52.

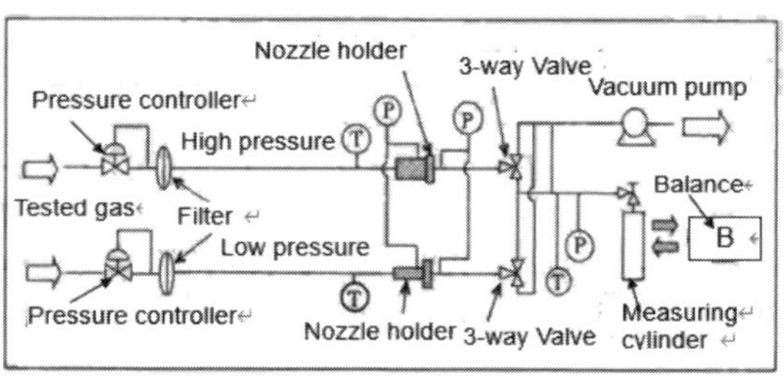

Fig.6.52 Weighing tank system (specified standard for small gas flow rate) (29)

Two test lines are used properly according to the flowrate range. The sonic nozzle, which is the flow meter to be calibrated, is attached to the holder shown in the figure. The procedure is

(1) Setting the pressure required for a calibration flowrate with a pressure control device, and turn on a vacuum pump.
(2) After the flow stabilizes, switching the flow to a measuring cylinder side with a three-way valve and let the flow into a measuring cylinder.
(3) After a set time has elapsed, close the three-way valve and finish a measurement.
(4) After that, measuring the change of weight of measuring cylinder with a balance.

Photo 6.28 Constant volume tank system

Photo 6.28 shows the gas calibration equipment for medium air flowrate in a constant volume tank. (6m long, 2m in diameter and 11m^3 in volume.) is set. The method of verifying sonic nozzles is as described above. (Related to Fig.6.50)

Overall, the contents of gas calibration equipment are as shown in the following Table 6.1.

Table 6.1: List of Gas Calibration Equipment of the National Institute of Advanced Industrial Science and Technology (As of January 2005) (29)

Classified name	Equipment name	Measuring target and range	Max. measuring ability (k=2)	Note
Small flowrate	Gas calibration equipment (Weighing tank System)	Sonic nozzle 0.005 to 180g/min (Dry air, N_2 gas) 0.1 to 200g/min (Argon gas) 0.05 to 60g/min) (Helium gas)	005 to 0.25%	Specified standard
Medium flowrate	Gas calibration equipment -Constant volume tank system-	Sonic nozzle 5-200m³/h	0.17%	Specified standard
Medium flowrate	Closed loop flow meter calibration equipment	Gas flow meter 5-1,000 m³/h 0.1-0.5 MPa	0.28 %	

Note: About the Sonic Nozzle

The sonic nozzle has same the structure as a normal nozzle (see Fig.5.20 in section 5.3.1), but it is used in a state where differential pressure is kept above the critical pressure determined by its shape, as shown in Fig.6.53.

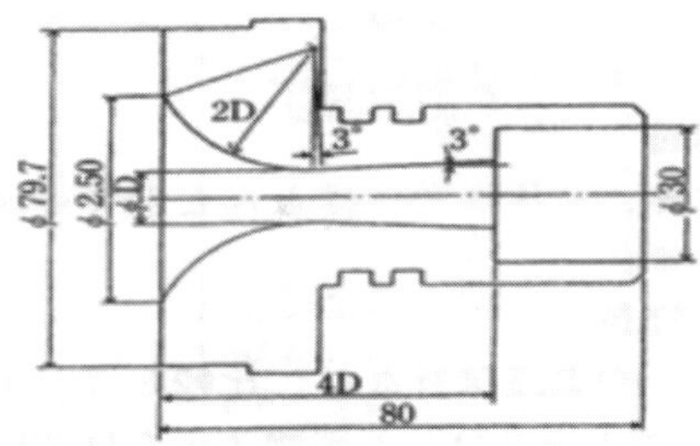

Fig.6.53 Example of Sonic Nozzle shape

The difference between the upstream pressure and the downstream pressure of the sonic nozzle (Fig.6.51-2) through which gas flow is gradually increased, and when the pressure ratio (downstream / upstream) becomes a certain value (critical pressure ratio) or less, the flow velocity of gas passing through throat portion (the narrowest part) is equal to the speed of sound. This is called the flow choking. That is, the constant flowrate can be obtained.

6.3.7 Japanese Water Flow Calibration Facility

AIST provides calibration services for liquid flow meters using seven calibration facilities as national standards. The calibration facilities for water flow cover the range between 2 L/h and 12,000 m^3/h with uncertainties between 0.039 % and 0.081 % are as in the following table 6.2.

Table 6.2 Facilities for water flowmeters

Facility	Main Equipment	Flow rate	Uncertainty ($k = 2$)	Remarks
Hi-Reff (High Reynolds number actual flow facility)	Reference flowmeters, Ball prover	750 m^3/h to 12, 000m^3/h	0.081 %	to 70 °C
Medium, Large	500kg,5t,50t Weighing tank and head tank	0.3 m^3/h to 3 000 m^3/h	0.042 % to 0.060 %	
Small	10 kg weighing tank	0.002 m^3/h to 1.2 m^3/h	0.039 %	

Hi-Reff (High Reynolds Number Actual Flow Facility)

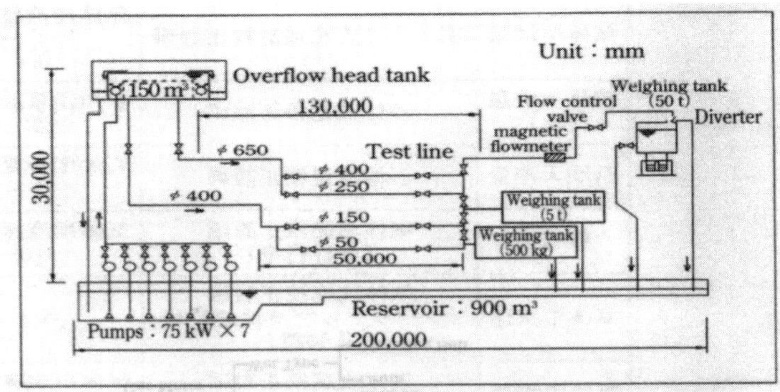

Fig. 6.54 **Water flow meter calibration equipment** (29)

To explain the very big (100m x 30m) Hi-Reff of Fig.6.54, an Overflow Head Tank is located 30m above the ground, and the water pumped up to the head tank flows down to the test tube 18.5m below by gravity. Hydrostatic flow within ± 0.5 mm of water level can be obtained. The accuracy of the flowmeter tested is obtained in comparison with a weighing tank.

General features

Flowrate	750 to 12,000 m³/h
Uncertainty	0.081 % (*k*=2)
Calibration Method	Master Meter Method
Temperature	up to 70 °C

■**Medium and Large**

General features

Flowrate	0.3 to 3,000 m^3/h
Uncertainty	0.042 to 0.060 % (k=2)
Calibration Method	Static and Gravimetric Method with Flying Start and Stop Method

■**Small**

General Features

Flowrate	0.002 to 1.2 m^3/h
Uncertainty	0.04 % (k=2)
Calibration Method	Static and Gravimetric Method with Flying Start and Stop Method

6.3.8 World Calibration Facilities for Flowmeters (29)

There are many excellent calibration facilities for flowmeters in several countries, so I would like to be list them.

Table 6.2 Several Calibration Facilities for Flowrate in the World [29]

Country	Name (place) Facility	Using fluid	Calibration method	Calibration condition
Netherlands	**NMi** Large flow rate calibration equipment	**Air**	Volumetric Standard devices : Bell prover : CVM: 4 units : Turbine meter of 3 units	Flowrate: 40 to 12,000 Nm^3/h Pressure: Atmospheric Uncertainty: 0.20 %
	Gasunie (Groningen) Research flow measuring device	**Natural gas**	Volumetric Standard devices : Flowmeters : Turbine meter of 4 units : CVM of 10 units, (of 12)	Flowrate: to 36,000 Nm^3/h Pressure: to 3.9MPa Uncertainty: 0.28 %
	Gasunie (Bergum) Calibration equipment	**Natural gas**	Volumetric Standard devices : Turbine meter of 5 units	Flowrate: to 14,4000 Nm^3/h Pressure: to 6.1MPa Uncertainty: 0.28 %
	Gasunie (Westerbork) Large flowrate device	**Natural gas**	Volumetric Standard device : Turbine meters 10 units	Flowrate: To 4,000,000 Nm^3/h Pressure: to 6.1MPa Uncertainty: 0.28 % Max. bore size: 1,600 mm
France	**Gaz de France**	**Natural gas**	PVTt method for Sonic nozzle Standard device : Sonic nozzle of 5 units	Flowrate: to 10,000 Nm^3/h Pressure: to 6MPa Outflow coefficient accuracy for tested sonic nozzle: ±0.25%
USA	**NIST**	**Air**	PVTt method Standard device : Constant volume tank	Flowrate: to5,000 Nm^3/h Pressure: to 0.86MPa Accuracy for mass flowrate: 0.11%
Germany	**PTB**	**Air**	Laser doppler method	Flowrate: 10 to 2,500 Nm^3/h

			Standard device : Laser doppler : Velocity meter (LDA)	Pressure: to 6MPa Accuracy for mass flowrate: 0.1%
England	**NEL** High pressure gas flowrate calibration device	**Air**	Mass method Standard device : Weighing vessel and scale	Flowrate: to 1800kg/h Pressure: to 5MPa Measuring accuracy for mass: ±0.01 kg
China	(Kaifeng) Gas flow rate standard device	**Air**	Weighing method Standard device : Bell provers of 3 unit	Flowrate 0.4 to 360 m³/h Pressure : 0.1 to 0.4MPa Accuracy ±0.5%
	(Kaifeng) Gas flow rate standard	**Air**	Standard device - Sonic nozzle 4 units	Flowrate 0 to 1,500 kg/h Max. pressure 0.25MPa [abs] Uncertainty of equipment: ±0.32%
	(Kaifeng) Large flow center for water	**Water**	Volumetric (Overhead tank Used, 20.5m high) Standard device - Standard tank 3 units	Flowrate Max. 18,000 m³/h Pressure - Head pressure Uncertainty: - Unknown
Korea	**KRISS** Gas calibration equipment	**Air**	Mass method and gyroscope method: Mainly for sonic nozzles	Flowrate 10 to 1240 m³/h Pressure - to 5MPa Uncertainty (Overall primary calibration) - ±0.05%

6.3.9 Natural Gas Situation in Europe

Until the 1950s, in Western Europe, the atmosphere had been covered with gas produced by “town gas” or by "coke gas," which were a by-product of the iron-making process, because of small gas factories located in the center of large cities. There were more than 700 "city gas" factories in France, causing inefficiency and air pollution.

In the United Kingdom, citizens were plagued by pea soup fog, which was said to be a London specialty. In 1962 a large-scale fog that affected the whole country for four days occurred. However, at the end of this, the switch from coal to natural gas began in 1967, and finally it was released from air pollution.

The conversion of Western Europe to natural gas had gradually arrived. Submarine gas was discovered in the Netherlands in the 1950s, followed by the North Sea in 1965, and then by Norway, Western Siberia and Algeria. In addition, technological advances had made it possible to compete with oil and coal in terms of price even if was sent by pipeline to a location several thousand kilometers away. Nowadays, pipelines of natural gas have been laid in Europe in every direction.

Under such circumstances, many flowmeters had come to be used for the transaction and management of natural gas pipelines between countries, and the Dutch Gasunie became the technical center of the transaction certification and accuracy maintenance.

Therefore, as shown in Fig.6.55, the traceability of natural gas calibration equipment from small flow rate to large flow rate was established by NMi and gas company Gasunie.

Westerbork of Stage 7 is the biggest equipment in the world.

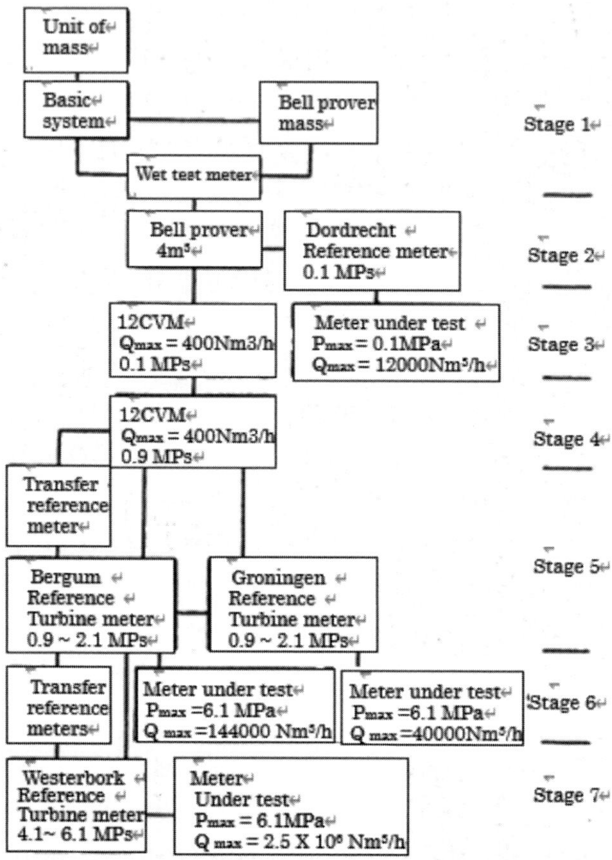

Fig. 6.55 Traceability for the equipment of NMi and Gasunie equipment [29]

Recently, a paper [70] related to this equipment was discovered, so I would like to introduce it.

"The Bernoulli laboratory ("Westerbork") in the Netherlands, jointly operated by Gasunie, NMi and KEMA, holds a record of over 30 years in calibrating very large gas meters at high pressure with natural gas.

(Text in the middle omitted). This facility was established in 1978, as the Bernoulli laboratory in Westerbork, the Netherlands, and provides research and calibration services with high pressure natural gas under flowing conditions.

The facility is among the largest in the world, capable of calibrating 30 inches gas meters with 120D straight upstream piping. Annually, more than 200 large meters are calibrated." (70)

Take a look at the aerial photographs (color) of this valuable facility.

Photo 6.29 shows the Bernoulli laboratory.

It is named after Bernoulli, who is famous for fluid dynamics!!

A building on the upper left of this photo is probably a calibration equipment room in which 10 units of reference turbine flowmeters are installed. And the straight pipes are thought to be extended into the building to the lower left.

A Blow down silencer allows testing while releasing natural gas into the atmosphere. There is no description of the site area, but it is probably 300m x 100m or more according to the photo. Figure 6.54 is a bird's-eye view of the Westerbork calibration equipment known so far, but this photo shows how amazing it is.

Photo 6.30 is a scene of 24-inch ultrasonic flowmeter being tested in this laboratory.

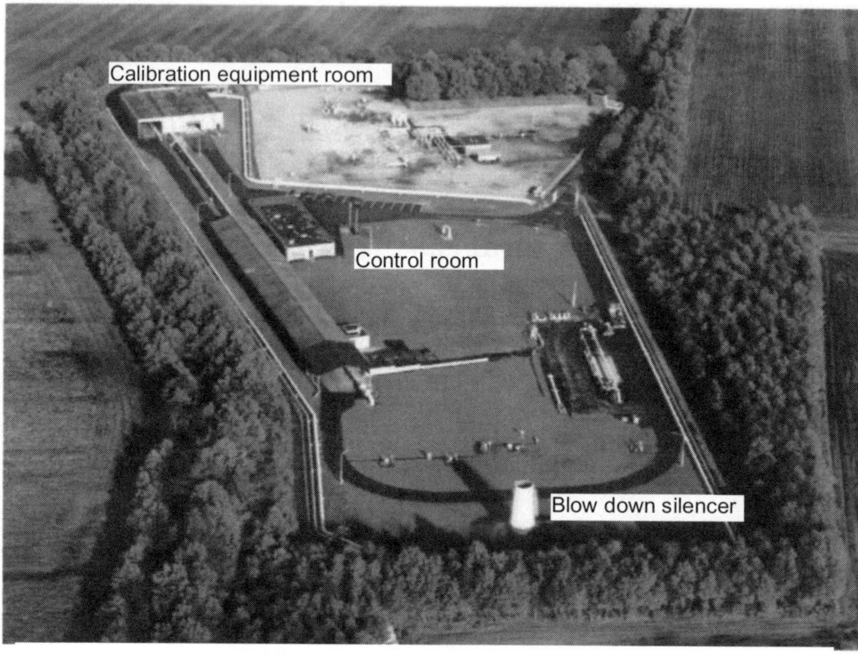

Photo 6.29 Bernoulli laboratory in Westerbork (70)

Photo 6.30 A scene of 24-inch ultrasonic flowmeter tested (70)

6.3.10. Traceability

Recently, the word traceability has become quite familiar. For example, housewives can understand where the tuna sold in the market was obtained and the distinction between Chinese and domestic vegetables.

Traceability is the ability to trace back to the source in this way, but it seems that traceability began to be mentioned in the measurement field around 1962. (54)

The American Standards Institute Conference was held in Boulder, USA, and "Traceability" was actively discussed.

In Japan, a few years later, the Traceability Research Committee started at the Institute of Electrical Engineers of Japan.

In addition, an article, "Traceability to measurement standards" was set up as a small special feature in "Measurement and control" in 1978. In it, the system was shown as in Fig.6.55 concerning the traceability of flowmeter at that time.

That is, what kind of standard was used to calibrate the end flowmeter? In addition, traceability in flowrate measurement is whether the standard is connected to the national standard of flowrate

measurement of quantities at the time of transactions or taxation performed in various places in social life, or for official certification are called Measuring Instruments to distinguish them from general measuring apparatuses. Today, most countries around the world have enacted

legislation on measuring instruments, initially verifying whether they were manufactured and repaired correctly, and controlling whether they are used in the correct condition. In Japan, that is the Measurement Law.
Performing certification and control under these laws are called Legal Metrology, and measuring instruments which perform the legal metrology are called Legal Measuring Instruments.

6.3.11 Trends of International Standards and Standardization for Flowrate Measurement

(1) OIML (International Organization of Legal Metrology)

It is desirable in various aspects, including trade, for this legal metrology and legal metrological instrument to maintain international harmony on measuring accuracy and display method of quantity. To realize these ideals, the International Organization of Legal Metrology (OIML), an intergovernmental organization, was established under the 1955 International Convention. The secretariat BIML is located in Paris, France.

(2) ISO (International Organization for Standardization)

On the other hand, the International Organization for Standardization was established in 1947. In English, but the abbreviation is ISO, not IOS. A reason is that ISO comes from the Greek word "isos" which means "equal", and it is also the origin of English isonomy (equality under the law)

and isometric (equal size), and they are equal, equality, etc., are abbreviated as "standard" and "standardization".
The 28th Technical Committee (TC28, petroleum products and lubricating oils) and the 30th Technical Committee (flowrate measurement in pipelines) are in charge of measuring the amount of fluid.

ISO is legally a non-governmental organization (NGO), but it is affiliated with representative standardization bodies (mostly public corporations' organization) in each country. It can be considered as an upper mechanism of JIS (Japanese Industrial Standards).
There are so many ISO standards for flow measurement that there is no time to list them.

6.3.12 History of Measurement Uncertainty (59)

Now, the measurement accuracy of a flowmeter is an important issue for the user. Therefore, the salesman who sells flowmeters almost always says, "This flowmeter has good accuracy." In addition, it is customary to write the accuracy ± 0.2% etc. in the catalogue. However, recently, it has come to be said that the quality of instruments or the quality of measurement should be expressed by "uncertainty".

Taking Japan as an example, a revision committee for JIS Z 8101 "Quality Control Terms" was held around 1980, and the positions of physical measurement and chemical analysis were a head-on conflict over the definitions of "accuracy", "correctness", "exactness", and "precision" from the beginning, but it was revised in 1981 as still in place today.

On the other hand, at ISO, TC29 "Application of statistical methods" (secretary country: France) started to move. This technical committee was preparing the draft of ISO 5725 "Accuracy and Precision of Measurement Results", but it is said that concerning the measurement condition of "dispersion" to evaluate the variation of measurement results, was in a serious conflict for more than 10 years with the United States (formerly NBS), which insisted on a form close to repeated conditions and Japan, which insisted on a form close to reproduced conditions.
At first, Japan was isolated and unsupported, but in the early 1980s, with the support of Canada, Britain, Germany, and France, it seems that the American way of thinking became the minority. Also, Japan abolished the negative concept of measurement "bias" and proposed the term "trueness", which received tremendous support.

After that, ISO TAG4 came to be published in 1992 with GUM (Guide to the Expression of Uncertainty in Measurement), an ISO guide that also incorporated Bayesian statistics along with IEC (International Electrotechnical Commission), OIML, and BIPM.

That is, it included a Type A evaluation based on statistical evidence and a Type B evaluation based on subjective estimation. However, this is still controversial and requires work to scientifically consider GUM itself and clarify its scope of application.

Although the author is not familiar with statistics, thinking about it carefully, it can be understood that error is ± 0.2%, but accuracy of ± 0.2% seems to be a strange expression. Should it be 99.8%? Also, "uncertainty" is a negative expression, so manufacturers do not want to use it. Why can it be expressed as "Certainty 99.8%" instead of "uncertainty 0.2%"? But it seems to come from statistics that a 100% true value cannot exist. That is, the probability for a true value that does not exist may be nonsense.

Nowadays, the expressions of accuracy are as follows from each company's catalogue.

Accuracy: ± 0.2% of reading

Maximum measurement error: ± 0.2% and so on.

6.4 Trends in New Technologies in Flowrate Measurement

The history of flow measurement is approaching the ending, but at the end I would be like to talk about the trends for new technologies in flow measurement and to end with future development expectations.

6.4.1 Hydrogen Flowrate Measurement [60][61]

As a measure against global warming, CO_2 reduction has been called for, and fuel cells have been rapidly attracting attention in recent years.

The fuel cell is a power generation system that reforms

The fuel cell is a power generation system that reforms fossil fuels such as petroleum, natural gas, and biogas to produce hydrogen, and chemically reacts hydrogen and oxygen to extract electricity and water. In addition, fuel cells are being developed in a wide range from power generation systems for hospitals etc. to usages for homes, automobiles and personal computers.

In December 2002, leasing sales of fuel cell vehicles (FC vehicles) began in Tokyo and a filling station for the purpose of verifying hydrogen supply to FC vehicles, which is a national project, has become necessary.

Tatsuno Mechatronics has developed a hydrogen filling machine using a mass flowmeter as shown in Photo 6.31. Filling pressure is as high as 25 MPa (250 atm) or 35 MPa

Photo 6.31 Tatsuno hydrogen filling machine

Photo 6.32 Coriolis flow meter for ultra-high pressure (provided by Oval)

(350 atm).

The flowmeter used for this is an Oval ultra-high pressure Coriolis flowmeter with the shape shown in Photo 6.32, which is currently used for hydrogen filling of hydrogen fuel cell vehicles at the Aichi Expo. Such ultra-high-pressure hydrogen can be measured with a flow meter that detects by force, but as is well known, hydrogen is the lightest gas molecule wise and it is difficult to measure the flowrate at low pressure, but as is also well known, to the contrary, since the specific heat of hydrogen is large, a thermal mass flow meter which is a thermal detection method is suitable. Photo 6.33 shows the appearance of a recent digital thermal mass flow controller (BRONKHOST made), which has a digital output with a flow control valve.

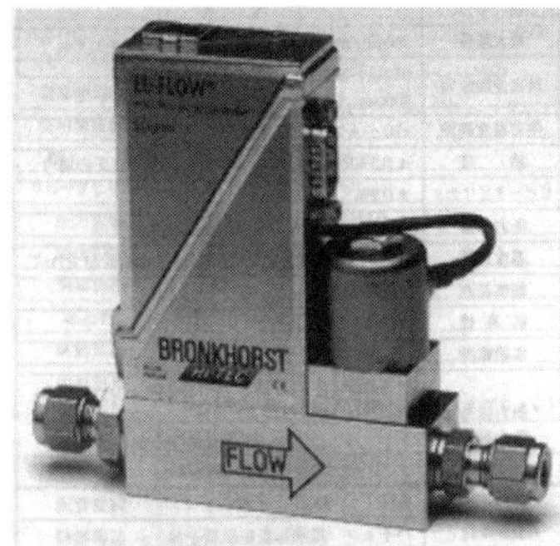

Photo 6.33 Digital thermal mass flow controller (61)

6.4.2 User-Friendly Flow Measurement

Recently, there has been a request from the user side to reduce the number of maintenance personnel for measuring instruments in the entire factory as much as possible. In addition, each flowmeter manufacturer has also provided user-friendly flow meters. Here are some of them.

(1) Electromagnetic Flowmeter with Electrode Adhesion Prediction Diagnosis Function (62)

Yokogawa Electric has been selling the ADMAG series of dual-frequency electromagnetic flowmeters, but announced that it has an electrode adhesion prediction diagnosis function added to it. This is the first time in the world for this. The adhesion state of insulator adhered to the electrode surface is constantly monitored and the adhesive level is displayed. The measurement principle is shown in Fig. 6.56.

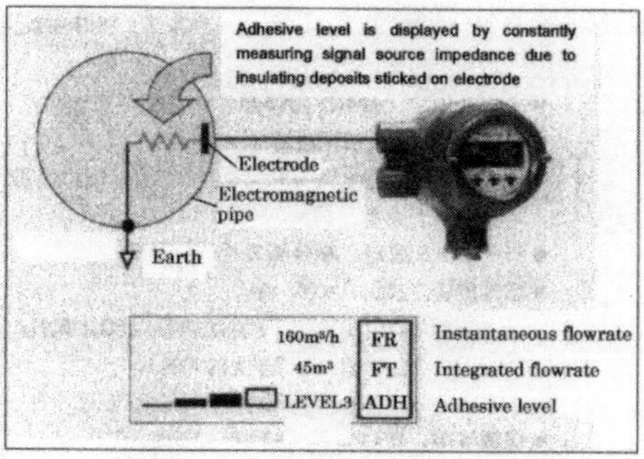

Fig.6.56 Diagnostic function of adhesive prediction on electrode (62)

Photo 6.32 shows the detachable electrode structure to facilitate maintenance. As a result, in the field where the adhesive fluid is measured, the electrode cleaning cycle is performed based on the experience so far, but it can be performed based on this display, enabling efficient maintenance.

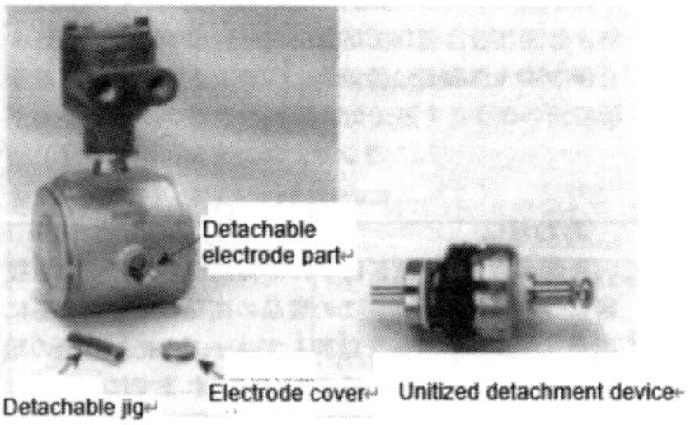

Photo 6.32 Detachable type Electromagnetic flowmeter (62)

(2) Maintenance Function of Coriolis Flowmeter [63]

Recent Coriolis flowmeters have the following diagnostic functions:

(1) Air bubbles are mixed in

(2) Measurement fluid adheres to the tube

(3) the effect of piping vibration

The weaknesses of the Coriolis flowmeter can be compensated by diagnostic functions.

By using the communication between the flow meter and a commercially available Windows PC., judgement of the situation can be helpful for maintenance. Photo 6.34 is an example of using a mobile personal computer (Oval Corp.).

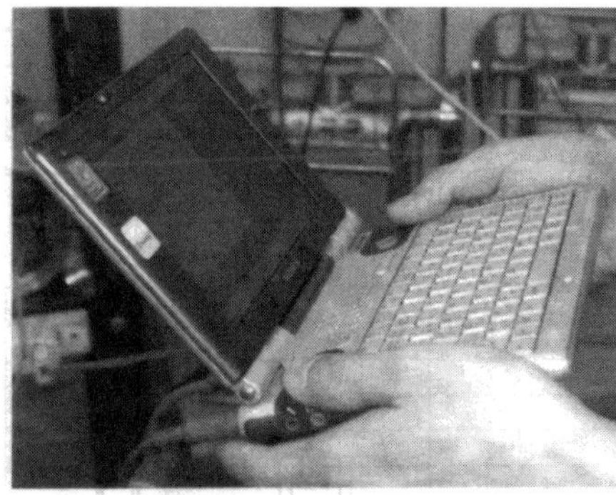

Photo 6.34 Diagnostic function by communication between flow meter and personal computer [63]

6.4.3 Review of Flowmeter

Not only flowmeters with a new principle are in the limelight, but there are many examples of flowmeters with an old principle that have been improved and improved in performance.

Such as the differential pressure type flowmeter that dates back to Bernoulli's theorem in 1740, and the multi-digital flowmeter of Tokyo Keiso Co., Ltd., and a V-cone meter

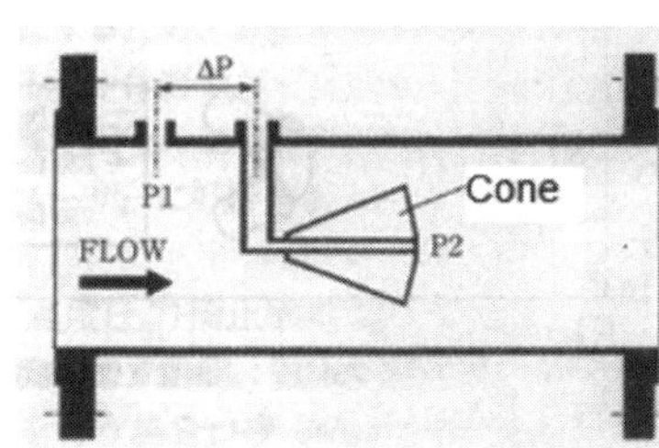

Fig. 6.57 V-cone flowmeter

shown in Fig.6.57 was developed by Macrometer Inc. of the United States. There is a merit in that a cone-shaped throttle for the contraction of area has the advantage that the influence of piping is small and the required straight pipe length before and after is less than that of the conventional orifice.

-That all-

At the End

This book is an English translation of the original that was published in Japanese 15 years ago.
The title is “Nagare-waza no siruku ro-do”, which means "Silk Road of Flow Measurements" in English.
This is a description of how the technology related to flow measurements that had begun in Western Europe was transmitted to Japan at the end of Asia. This journey is based on the name "Silk Road of Flow Measurements", imitating the famous Silk Road
It started with a Nilometer that measured the water level of the Nile in 4,000 BC, and I intended to describe in detail when old and recent technologies related to flow measurement had been transmitted to Japan over 6,000 years. In addition, I determined recently to make an English version because many readers would like to read this. There is no book with such content in the world!

Now, we live on the planet Earth, which orbits the stellar Sun in the galaxies of the universe. Looking back on the

history of the Earth,

- the Earth is 4.5 billion years old and creatures appeared 2.4 billion years ago.
  (For reference, the universe was born 13.8 billion years ago according to the most recent research.)
- The Ice Age came at least three times and only a handful of Aerobic Organisms survived.
- The Ice Age was ended by volcanic eruption. Activated cyanobacteria increased oxygen, as a result, the number of organisms using oxygen increased.
- However, 440 million years ago, 45% of the creatures were killed by gamma-ray bursts, which were said by some to be beam weapons from space. After that, acid rain continued and many creatures became extinct.
- It is said that in the Helm period 250 million years ago, an unprecedented volcanic eruption destroyed 90% of marine lives and 70% of land lives. However, large dinosaurs and mammals appeared.
- However, 65 million years ago, an asteroid (the size of Mt. Everest) collided with the earth, the dinosaurs that were enjoying the era became extinct. But this was good for mammals and led to the emergence of humankind.
- Mankind appeared in Africa only about 7 million years ago.

Compared to the history of space and the earth, 6,000 years is a very short time, but I think it is a wonderful progress. However, looking back on the history of the earth, there have been many mass extinctions, and the earth is by no means a safe place. Preventing mass extinction is a consultation that cannot be done by humankind, and we can only wish for good luck in the future.

Long live the earth! Good luck with flow measurement! Finally,

I dedicate this book to my late wife (March 16, 2020)

Reference

(1) "Greater East Asia Co-prosperity Sunk in Leyte Island", NHK interview team

(2) "100 Years History of Nippon Oil", edited by Nippon Oil Co., Ltd., 1988

(3) Mitsutomo Yuasa "Chronology of Science and Culture" Chuokoron-sha

(4) McMillan "World Historical Statistics (1) (2) (3) 1750-1993" Toyo Shorin

(5) Nippon Oil Mitsubishi Co., Ltd. "Oil Handbook 2000", Fuel Oil and Fat Newspaper

(6) "History of Japanese Industry 2" supervised by Hiromi Arisawa, Nihon Keizai Shimbun

(7) Meeting for measurement Meeting "Measurement J vol.1, N0.1, Corona Publishing Co., Ltd., 1950,12,25

(8) "75 Years History of Yamatake Honeywell " Yamatake Honeywell, 1982

(9) "Devoting measuring instruments -50 Years of Yokogawa Electric " Yokogawa Electric Mfg. Co., Ltd., 1965

(10) "Hokushin Electric Works: 50th Anniversary-" Diamond, 1969

(11) "Fifty Year History of Tokico" Tokico Co., Ltd., 1967

(12) Jun Kashima: "Selfless Immediately Forty Years" Oval Gear Engineering Co., Ltd., 1988

(13) Toshio Hirano: "Recollection of the late President Murakami" Oval News N0.11, 1953, 07, 10

(14) Meeting for measurement Meeting "Measurement", vol.1, N0.2,1952,01,01

(15) "Special Issue for Oil (Vol. 2)" Oval News, N0.89,1961

(16) "50 Years of Tokico" Tokico, 1988

(17) " Roots Flowmeter Combined Cycloid 45 ", Tokico Review, N0.29, 1970

(18) "Oval News", catalogs, etc.

(19) "Flow Measurement A to Z", edited by Japan Metrology Equipment Manufacturers Association

(20) Kazuto Kenno: "About the Electromagnetic Flowmeter", measurement vol.3, N0.4, 1953
(21)) H. G. Elrod "An Investigation of Electromagnetic Flowmeters" Transaction of A.S.M.E, May 1952
(22) Kenno Kazuto" Current Status of Electromagnetic Flowmeters" Measurement vol.5, N0.8, 1955
(23) "Houchin Electric Works", edited by Houchin Electric Works Limited, 1969
(24) Shoo Matsushiro: "Revised Flow Rate", Corona Publishing Co., Ltd.
(25) V. Cushing "Electromagnetic Flowmeter" Flow 1974
(26) J. L. MCSHANE "Ultrasonic Flowmeter" Flow, Instrument Society of America, 1974
(27) J. Kritz "Ultrasonic Flowmeter ¨ Instrument &Automation,Nov.1955
(28) Yoshiaki Yamamoto: "Ultrasonic Flowmeter" Measurement and Control, vol.18, N0.5,1979
(29) "Flow Measurement A to Z", edited by Japan Metrology Equipment Manufacturers Association, 2005
(30) Japan Panametrics Catalog, "DigitalFlowTMGC868"
(31) "Multi Window" Keiso 2004, January issue
(32) Keiso 2004, January issue
(32) Takashi Nakata: "In a dialogue in MESSAGE", SPECIALTY N0.8, Oval, 1990,01
(33) Theodore von Kalman, Translated by Ichiro Tani: "Theory of Flight", 1955
(34) Theodore von Karman "Der Mechanizmus des Flussig-keits und Luftwiderstandes" Gottingen Nacher. 1911, 507
(35) D. F. White, A. E. Rodely "The Vortex Shedding Flowmeter", Flow Instrument Society of America,1974
(36) "YEWFLO Vortex Flowmeter" Measurement and Control, vol.18, N0.10,1979
(37) Ogawa and Matsubara: "New Type of Vortex Shedding Flowmeter" FLUCOME TOKYO '85, 1985
(38) Masaki Takamoto et al. "Development of a Standard Vortex Shedding Flowmeter" FLOMEKO '94, 1994,6
(39) James M Benson "Survey of Thermal Devices for Measurement Flow" Flow, 1.S.A, 1974
(40) Raymond T. Doyle "Metering Gas with a Heated Thermopile" Instrument &Automation, 197.12
(41) J. Yoder "How the Technology Came to be Innovators" Flow Control, 2004,3
(42) Ogawa Scam, "Super Study for Flow Measurement Technique 4th, Coriolis Flow Meter", Automation, vo144, N0.8, 1998.8

(43) JIS B 7555.2003 "Flowrate Measurement Method by Coriolis Meter (Mass Flowrate, Density and Volume Flowrate measurement), explained "
(44) Water Meters" American Water Works Association, 1972
(45) R. E. Thompson "Turbine Flowmeter Performance Model"
(46) Hiroyuki Amemori, "Piping for Potter Meter Installation", Tokiko Review, N0.41, 1973.6
(47) Yutaka Ogawa, "Super Study Technique of Flow Measurement-5th Turbine Flow Meter", Automation, vol.44, N0.11, 1999.11
(48) Yoshio Tomita, "Instrumentation Engineer: Its History and Expectations for the Future", Measurement and Control, vol.36, N0.2, 1997.2
(49) Yoshio Tsukada, "Globalization and its Memories", Measurement Technology vol.31.N0.5,2003.4
(50) Kesakatsu Koizumi, "History of Weights and Measures", edited by Central Metrology Institute
(51) R. E. Speckmann "Proving High-Capadty Pipeline Meters", The oil and Gas Journal, 1954,8
(52) API2531 "Mechanica1 Displacement Meter Provers"
(50) Kesakatsu Koizumi, "History of Weights and Measures", edited by Central Metrology Institute
(54) "Traceability to Measurement Standards", Measurement and Control, vol.17, N0.3, 1978
(55) Kazuo Suzuki, "Fluid Measuring Equipment Industry", Measurement and Control, vol.17, N0.3,1978
(56) Koji Hosaka. "Trends in International Standards and Standardization Regarding Flowrate Measurement"
(57) Yutaka Ogawa, "International Standards for Flowmeters and Calibration Equipment for Flowmeters ", Automation, vo1.45, N0.3,2003.3
(58) "Positive-Displacement Meter for Motor Fuels", Engineering, 1938.10.28
(59) Hirokazu Tsubaki, "Uncertainty of GUM-Advantages of International Standardization and Various Problems" Workshop: New Technology of Vibration / Shock Measurement, Basic Knowledge of International Standard Activities" Japan Society of Mechanical Engineers, 2004.11.18
(60) Akifumi Kanamori, "Hydrogen Filling Machine and Infrastructure" Measurement Technology, vol.32, N0.1,2004
(61) Katsuo Misumi, "Measurement of Hydrogen Flow Rate", Measurement Technology, vol.32, N0.1,2004
(62) Hironobu Ota, "New Electromagnetic Flowmeter ADMAG AXF Series", Measurement Technology, vol.31, N0.12,2003
(63) Osamu Ono, "Maintenance Function of Coriolis Mass Flowmeter", Measurement technology, vol.33, N0.7, 2005

(64) *"Review of Flowmeter (1) (2)" Measurement technology, vo1.32, N0.4,5,2004*
(65) *"Measurement Standards Handbook" National Institute of Advanced Industrial Science and Technology*
(66) *C. Halse11 "Design of a Pvover for Mass Measurement of Gas" West Virginia University, 1959*
(67) *E.C.Evans "Gas Mass Flow Rate Measurement to 0.1%"Instrumentation &Control systems, vol.38, 1 965*
(68) *Masaki Takamoto "Standard of Flowrate and Flow Velocity" Measurement and Control, Vol. 33, No. 9, 1994*
(69) *Ishibashi, Takamoto "Pressurization Test of Ultra-precision Sonic Nozzle", Proceedings of The Society of Instrument and Control Engineers, Japan, 36-1 / 9 (2000)*
(70) *Calibration Errors of Ultrasonic Meters in the Bernoulli Laboratory Due to Non-isothermal Flow Conditions. Authors: Aernout van den Heuvel, KEMA (previously: Gasunie Engineering & Technology), others.*

CHRONOLOGY

Christian Era		Flow measurement technology	History of science and
Century	Era	(History of fluid technology included)	technology (Others)
BC30	BC 3000	- Ancient Egyptians measure the flowrate of the Nile by water level (the origin of Nilometer)	- (Emergence of Egyptian civilization around 5000), - Copper smelting around 3000
BC29		- The oldest dam constructed in Egypt Khufu era (2900 to 2877)	- (2800 The largest pyramid constructed)
BC26	BC 2500	- Copper water pipe (diameter of 40 mm, thickness of 10 mm, and length of 400 m) used in Egypt, - The Invention of waterwheel	
BC20		-Use of clay pipe culverts in Egypt (excavated at the Knossos palace)	
BC19 to BC8			- (1193-1184 Trojan War) -1400 Cast iron method developed, - (Birth of the Greek poet Homer around 900)
BC7	700 700 620	- Weighing water clock in Babylonia - Waterworks in Jerusalem - Water supply system in Nineveh	(753 Rome established) (606 Assyrian kingdom destroyed)
BC6	585 to 520	- Water works started in Babylon, - Qanat were built during the Achaemenid Empire of Persia, - Construction of an aqueduct in Greece (Aqueduct of Samos), - Construction of a sewer system called " Cloaca Maxima " in Rome	-Pythagoras (571 to 497) Greek mathematics, -Buddha (565 to 485), -Confucius (552 to 476),
BC4	From 320	- Construction of waterworks in Rome (4 such as the Appian Aqueduct)	- (331 Unification by the Great Alexander), - Geometry by Euclidean (330 to 275)
BC3 to BC2	180 159 135 to	- Construction of a siphon aqueduct in Greece (Pergamon Aqueduct), - Water clocks used in Rome, - Heron of Greece invents formula for calculating flowrate, - Appearance of the Persian water wheel (Sakiya) and drum water wheel - Archimedes screw water pump	- (146 Carthage and Greece fall) -Heron (135-75) invented Siphon, digestive pumps, steam reaction turbines, hydraulic organs, etc. - Archimedes (287-212) principle
BC1		- 3 more Roman aqueducts, (Because of Rome population up to 500,000 due to swelling in humans) - These became totally11 later.	- (Golden Age of Roman Culture), -(44 Caesar stabbed to death, Cleopatra appears)
1 to 13		- Goguryeo Doncyou conveys a waterwheel (Tengai) to Japan. - Use of horizontal windmills in Persia - Naka no Oue prince builds a Roukoku (water clock) and establishes the time signal system. - Windmill culture in China (Song Dynasty) from the 10 to 13 century - Windmills appear in the 12th Century England and France.	- (476 Western Roman Empire fell due to the Germanic tribes) - (6th to 8th century Dark Ages of the Eastern Roman Empire) -(645 Reform of Taika in Jap.) - 668 Japan, Echigo coal and petroleum discovered - (1192 Kamakura Shogunate starts)

Christian Era		Flow measurement technology	History of science and
Century	Era	(Including history of fluid technology)	technology (Others)
13	1290	-Japan's oldest "Uji water pump"	-(11 to 13th century: Crusader big expedition) -(13th century: The peak of scholastic philosophy) -1267 Roger Bacon: Experimental Science -(1299 Marco Polo: Memoirs of the East)
14		-In the 1350s in Japan, the waterwheels of Ujikawase and Oui existed.	-1302 Creation of a nautical compass -(1318 Dante's Divine Comedy) -(1352 Boccaccio' Decameron) -1370 Practical Gear clock by De Vic (Germany)
15	 1484 1490	-A lot of use of pumped-storage wind turbines in the Dutch sub-sea zone. -Leonardo da Vinci (1452 to 1519), invention of anemometer and hygrometer -Da Vinci: The observation of swirling water currents, sketches (Pioneer of modern science and technology)	-1450 Gutenberg: typography -(1453 The Eastern Roman Empire fell) -Da Vinci's painting: "Mona Lisa" -(1492 Columbus discovers a new continent) -Michelangelo: Sculpture, painting, architecture) -(Raffaello : Picture)
16	1500s	-Da Vinci discovered the sound transmissibility of water -Galileo (Italy, 1564 to 1642) said that the height of pump suction is 9m.	-Scientific renaissance era - Genius scientists such as Copernicus, Galileo, Kepler, etc. Gilbert appears one after another -1543 Guns introduced to Japan. -(1549 Xavier came to Japan)
17	1642 1643 1648 1652 1652 1654 1660	-Pascal (France, 1623-1662): devised the first calculator. Torricelli (Italian, 1608-1647) Torricelli measured the atmospheric pressure to be 760 mm of mercury. Invention of the mercury barometer. -Pascal, Observation of atmospheric pressure change with height -Pascal discovered Pascal's principle. -Guericke (Germany, 1602-1686): Invention of air pump -Guericke: Hemisphere experiment at Magdeburg --Boyle (British, 1627 to 1691): Boyle's law on gases	-(1600 Battle of Sekigahara) -(1602 Shakespeare, Hamlet) -1638 Galileo "Mechanical dialogue" -(1639 Iemitsu: National isolation Order) -1651 Experimental academy established (Italy) -1660 Hooke: the law of elasticity

Christian Era		Flow measurement technology (Including history of fluid technology)	History of science and technology (Others)
Century	Era		
17	1683	-Expansion of Watermill power factories in 17th Century Europe -Anemometer of Hooke (British, 1635 to 1727): -The Hooke announced open channel flowmeter -In the 17th century, Chen of the Qing Dynasty of China controlled flood by the water measurement method. -Newton (British, 1642 to 1727) announced the viscosity law rule	-1669 Newton invented calculus -1680 Leibniz (Germany, 1646 to 1716) invented the differential method. -(1685 Bach was born) -1686 Leibniz invented the integral method -1687 Newton, in "Principia" universal gravitation, Newton's first and second laws announced.
18	1728	-Pitot (France, 1695 to 1771) invented Pitot Tube	-1705 Newcomen (UK) improved steam engine -1740 Nollet (France) Experiment of static electricity.
	1740 1761 1766 1772 1783 1783 1787 1788 1790 1790 1794	-Bernoulli (Switzerland, 1700 to 1782) established Bernoulli's theorem -Euler (Switzerland, 1707 to 1783), establishment of fluid motion equations -Cavendish (UK), discovery of hydrogen -Rutherford (UK), discovery of Nitrogen -Scheele (Sweden), discovery of oxygen -Lavoisier asked Meunier for gas meter production. -Charles (France, 1746 to 1823), Hydrogen balloon invented -Charles announced Charles's Law of Gas expansion -Lavoisier established New View of Elements (Chemical Revolution) -Talleyrand (France) suggested the metric unit to parliament -Reinhard Woltmann (Germany) devised and created wheel-type flowmeter, -Lavoisier (France1743-1794): the father of modern chemistry-contributed to the accurate measurement of water density by the metric system, was put to death on the scaffold by guillotine	-1765 Watt (British): Steam Warfare -1768 Gennai Hiraga manufactured a thermometer. -1776 Gennai produced Elekiteru. -(1776 The United States Declaration of Independence,) -1780 Galvani (Italy) discovered Animal electricity -(1781 Construction of Kant's Philosophy) -1784 Argan (Switzerland), invented a lamp. -1788 Watt designs a centrifugal governor - (The beginning of automatic control) -1789 The French Revolution broke out) -1789 Lavoisier Published "Chemistry Handbook" -(1793 Marie Antoinette sentenced to death by guillotine -French Terrorism-)

Christian Era		Flow measurement technology (Including history of fluid technology)	History of science and technology (Others)
Century	Era		
18	1797	-Venturi (Italy, 1746-1822) presented the venturi meter.	-1794 Ecole Polytechnic established in France -(1796 Napoleon expedited to Italy) -1799 Volta (Italy) discovered a battery.
19	1802	-Gay-Lussac (French): the law of gas expansion	-1807 Ignition of gas lamped in London
	1819	-Jon Malam improved Clegg's wet gas meter.	-1812 Babbage (British) advocated a basic computer function.
	1825	-Crosley builds bucket meters	-1814 Stevenson (British) succeeded the trial run of steam locomotive.
	1832	-Faraday attempted to measure the flow of the Thames using the principle of electromagnetic induction	-1820 Oersted (Denmark), discovered the magnetic effect of electric current.
	1835	-Coriolis (France, 1792-1843), discovered the power of Coriolis	-1823 Siebold visited Japan
	1836	-Nariaki Tokugawa compiled "Ungeikisan" and classified pumping vehicles into 5 types.	-1825 World's first railway in England
	1837	-Kakuan Udagawa introduces the float-type hydrometer in the "Semikaisou"	-1831 Faraday (British) discovered electro-magnetic induction phenomenon.
	1840	-Hoyden devises a hook gauge for liquid level detection	-1844 Practical use of the Morse telegraph
	1842	-Doppler (Austria 1803-1853, Doppler effect of sound waves discovered.	
	1845	-Stokes (British, 1819-1903), derived the Navier-Stokes equation	-1847 Young (British) Succeeded in producing petroleum by dry distillation of coal.
	1847	-Frode developed the flow nozzle.	
	1850	-Thomas Glover completed dual-membrane gas meter.	-1847 Siemens Co. was established.
	1851	-Dunn developed the first wing wheel water meter.	-(1847 Economic depression)
	1854	-Kawamoto Koumin explained the steam engine in "Ensei-kiki-jyutu"	-(1853 US Ambassador Perry comes to Uraga) -1859 Drake (US) excavates successfully oil well
		- . -	-(1867Japan, restored the imperial rule)
	1868	Chameroy patented an area flow meter (float type).	-1870 Rockefeller (U.S.) established Standard Oil
	1873	-Deacon developed a disk-type water meter (with a recorder)	-1872 Shinbashi-Yokohama railroad opened
	1874	Reynolds (British, 1842 to 1912) proposed the Reynolds number, which expresses the flow state.	-1879 Edison's invented a carbon wire type light bulb.

Christian Era		Flow measurement technology (Including history of fluid technology)	History of science and technology (Others)
Century	**Era**		
19	1884	-National Meter Co. developed a rotating piston water meter.	
	1886	-Japan joined the meter convention.	-1886 Daimler and Benz produced gasoline engine cars.
	1887	-Herschel (U.S., 1842-1930) builds a venturi meter.	
		-Denton (US) introduced papers of Saybolt(US) viscometer, Perkin viscometer, etc.	-1888 Hisahiro Naito founded Japan Oil Co. -1888 The completion of Kodak camera
		-Palmer (British) laid Japan's first water supply in Yokohama	
	1891	-Japan enacted the Weights and Measures Law (official recognition of the metric system)	
	1893	-Enactment of Syakkanhou based on the prototype of the metric meter	-1893 The invention of diesel engine - (1894 Sino-Japanese War)
	1896	-Meinecke designed and manufactured disc meters.	-1897 Germany invented the cathode-ray tube
	1898	-Completion of waterworks in Tokyo	
20	1904	-Prandtl (Germany, 1875 to 1953) published the boundary layer theory -Jyumonji Daigen completed domestic gas meters.	-1903 Wright brothers' success the first airplane - 1903 Ford Motor Company established -(1904 Russo-Japanese War broken out) -1904 Fleming invented the diode vacuum tube
	1906	-Hypersberg Co.produced the rotary piston meter.	-The electrification of Railway began.
	1906	-Jyumonji Daigen succeeded in developing domestically produced Water Meters	-1905 Einstein: The Special Theory of Relativity
	1909	-Gale developed the Sarco-meter (Steam Flowmeter)	-1906 Yamaguchi Shokai established. (Predecessor of Yamatake)
	1909	-Hodgson Developed D, D/2 Combination Tap Orifice	
	1911	-Thomas (US) developed a Thermal Flowmeter	-1910 Tokyo Gas Electric Industry established (Predecessor of Tokico)
	1911	-Karman (Germany,1881-1963) published Karman vortex theory	-(1914 World War I break out)
	1913	-Reebok announces the flowrate formula for full width weir	-1915 Electric Instrument Laboratory established (Predecessor of Yokogawa Electric)
	1914	-King (UK), Theoretical analysis of Wind Speed Measurements with Kata thermometers	
	1916	-Japan began the certification of gas meters	- (1917 Russian Revolution) -1917 Whittaker: Vibration engineering theory
	1919	-Kunze applied for a patent of wind speed measurement by ultrasonic	-1918 Hokushin Electric Works established.
	1919	-Visible gas station appeared in Tokyo	
	1920	-Reed & Northrup Inc. (USA) released the servo-type differential pressure detector.	

Christian Era		Flow measurement technology (Including history of fluid technology)	History of science and technology (Others)
Century	Era		
20	1922 1923 1926 1928 1928 1930 1930	-Williams applied Electromagnetic Flowmeters to circular pipes. -Hodgson develops the Shunt-type Flowmeter using orifice and impeller -Karman visited Japan at the invitation of Kawanishi Aircraft -New Jersey Meter commercialized the Area-type Flowmeter similar to current model. -Japan begins the certification of Water Meters -Around 1930, Askania measuring mechanism: Redox bell-shaped measuring mechanism, European Ring Balance Meter, and Cochrane Integrator were used to detect differential pressure. -Around 1930 Bassler Meter, Pittsburgh Meter, and Badger Meter appeared as reciprocating piston type Positive Displacement Oil Meters in the United States, and were also widely used as meters for gasoline weighing machines. -Around 1930 Roots Meters appeared as gas meters	-(1921 Washington Conference : Four Power Treaty -(1922 the Union of Soviet Socialist Republics established. -1923 Great Kanto Earthquake occurred in Japan. -1925 Heisenberg's quantum mechanics -1925 First practical television appeared. -1927 Lindbergh completed non-stop transatlantic flight -1929 Zeppelin rounded the world, and finally burned out。 -(1930 Germany, Nazi Party power increases)
	1933 1934 1936 1937 1938 1939 1940 1940	-Bopp & Reuters (Germany) developed Oval Meter -Japan began the Certification of Gasoline Meters -Colin applied electromagnetic flowmeters to blood flow measurement -Fischer & Porter mass-produced rotameters (area type) -Potter developed the electromagnetic pickup type turbine meter. -Tatsuno Seisakusho manufactured Japan's first meter-type gasoline meters. -Around 1940 sliding vane type meters appeared -The ASME flowmeter technical committee implemented the performance tests of various petroleum meters by reference tanks	-(1933 US President: Roosevelt) -(1933 Germany formed Hitler's dictatorship) -1935 Hideki Yukawa, introduced the meson concept. 1938 Invention of nylon -(1939 World War II began)
	1943	-Barton Corporation (US) developed a burst-proof bellows meter with temperature compensation.	-1943 Practical use of radar in UK and US -1944 Germany's V1 rocket appeared -1944 Oldenburg: automatic control theory

Christian Era		Flow measurement technology (Including history of fluid technology)	History of science and technology (Others)
Century	Era		
		-1945 Mauchly completed the world's first electronic calculator (ENIAC)	-1945 US created atomic bombs -(1945 World War II ends) -1946 Reconstruction policy of Japan "Incline Production System") implemented.
	1947	-International Organization for Standardization (ISO) established	-(1948 Gandhi in India assassinated) -(1948 Beginning of the Soviet blockade of Berlin) -1948 Shockley develops transistor
	1949	-The appearance of ultrasonic sing-around method	-1949 Hideki Yukawa won the Nobel Prize in Physics -1949 The partnerships between Japanese and foreign companies succussed.
	1949	- Oval meters and Roots meters for liquids made in Japan appeared	-1949 Oval Kiki Kogyo established
20	1950	-Around 1950 Rockwell's "Roto-cycle Meter" and Brodie's "Brodie Meter" appeared. -System: The Era of Pneumatic PID Controllers	-1949 UK put jet airliners into practical use. - (1949 Mao Zedong proclaimed the People's Republic of China (PRC) - (1950 Korean War broke out)
	1951	-The birth of Measurement Law in Japan	-(1951 Japan signs the San Francisco Peace Treaty)
	1951	- Mittelmann commercialized Electromagnetic Flowmeter -Foxborough and Endress+Hauser Introduce Electromagnetic Flowmeters	-1951 Establishment of the Measurement Council, publication of "Measurement" published, -1951 Bunji Okazaki built Japan's first electronic computer
	1954	-U.S. Shell Oil announced "Miracle Mile" proving equipment	-1952 Japan began television broadcasting
	1955	-Maxon presented the first Ultrasonic Flowmeter for liquids -International Legal Metrology Organization (O1ML) established	-1952 U.S. began production of junction-type transistors - (1953 Stalin died) -1954 Soviet Union: the world's first nuclear power plant
	1957	- Hokushin Denki announced Japan's first electromagnetic flowmeter -Raymond introduced thermocouple thermal flowmeter	-1957 Leona Ezaki invented the Ezaki diode -(1959 Castro: Cuban: Revolution)
	1960	-In the 1960s, computers began to be used for process control	

Christian Era		Flow measurement technology (Including history of fluid technology)	History of science and technology (Others)
Century	Era		
	1961 1962 1963 1964 1967 1968 1970 1970	-Japan joined OIML -Discussions on traceability in the field of measurement begin at the National Institute of Standards Conference "Pipe prover (reference volume pipe)" in AP12531 established. -Shipin (US) Obtains a patent for Coriolis flowmeter. -The development of DDC (direct digital control) advances in Japan. -Rodely and White develop Industrial Vortex Flowmeter -Benson introduces a Separate type Thermal Mass Flow. -API Standard 2534 for Turbine Flowmeter published	-1962 The announcement of Josephson effect -1963 The assassination of President Kennedy is played in the first space TV broadcast between Japan and the United States. -1964 Sharp in Japan starts the production of transistor calculators. -1964 Tokyo Olympics held - (1965 European Community EC established) -(1967 Chinese Cultural Revolution) -1968 Texas Instrument announces prototype MOS LSI. -1969 Apollo 11 lands the moon first. Two men land on the moon. -(1970 Osaka World Exposition held for the first time in Asia.
20	1971 1972 1974 1975 1980	-The signal of the electronic control unit is unified to 4 to 20mA. -A Doppler Ultrasonic Flowmeter is developed. -J.E. Smith develops the Coriolis Flowmeter and establishes Micro Motion, Inc. -Distributed Control Systems (DCS) emerge nearly simultaneously in the U.S. and Japan, that is, Yokogawa Electric: CENTUM, and Honeywell (Yamatake): TDCS2000. -Friend's Orin and Kurz found Sierra Instrument and Kurz instrument for Thermal Mass Flow Meter respectively. -JIS Z8765-1980 "Method for measuring flow rate by Turbine Flowmeter" established	-1971 U.S. announces microcomputer -1972 1BM Corp. announces Floppy Disk -(1973 End of the Vietnam War) -(1973 Japan, first oil crisis) -1976 Supersonic passenger plane Concorde goes into service - (Japan and US, The Lockheed bribery scandals comes to light) -(Zhou Enlai and Mao Zedong died one after another)
	1982 1987	-Micromotion commercialized two U-tube type Coriolis Flowmeter. -Endress +Hauser announced two-straight-tube Coriolis Flowmeter	-1981 The first flight of Space Shuttle Successes -1981 Kenichi Fukui: Nobel Prize in Chemistry -1981 The first group of orphans left behind in China came from China to Japan.

INDEX

D

H

I

J

K

U

V

W

X

Y

Z

PERSONAL EDITION

ABOUT THE AUTHOR

2023-04-19

I was born in Yokohama, Japan on July 10, 1938. The eldest of four siblings: Yuwakiti, Chieko and Ogawa.

In April 1945, I entered the National Elementary School in Dalian, China at the end of the Pacific War, and finally graduated from the Yokohama National University, Department of Mechanical Engineering. After that, I worked for the Oval Corporation on the design and development of flowmeters for about 40 years. I became a director and head of the research division in 1990. In 1977, I received the Metrology Merit Award from the Minister of International Trade and Industry Japan, and retired from the company in 2004.

APEX
TP
9781642733228
76902033-11